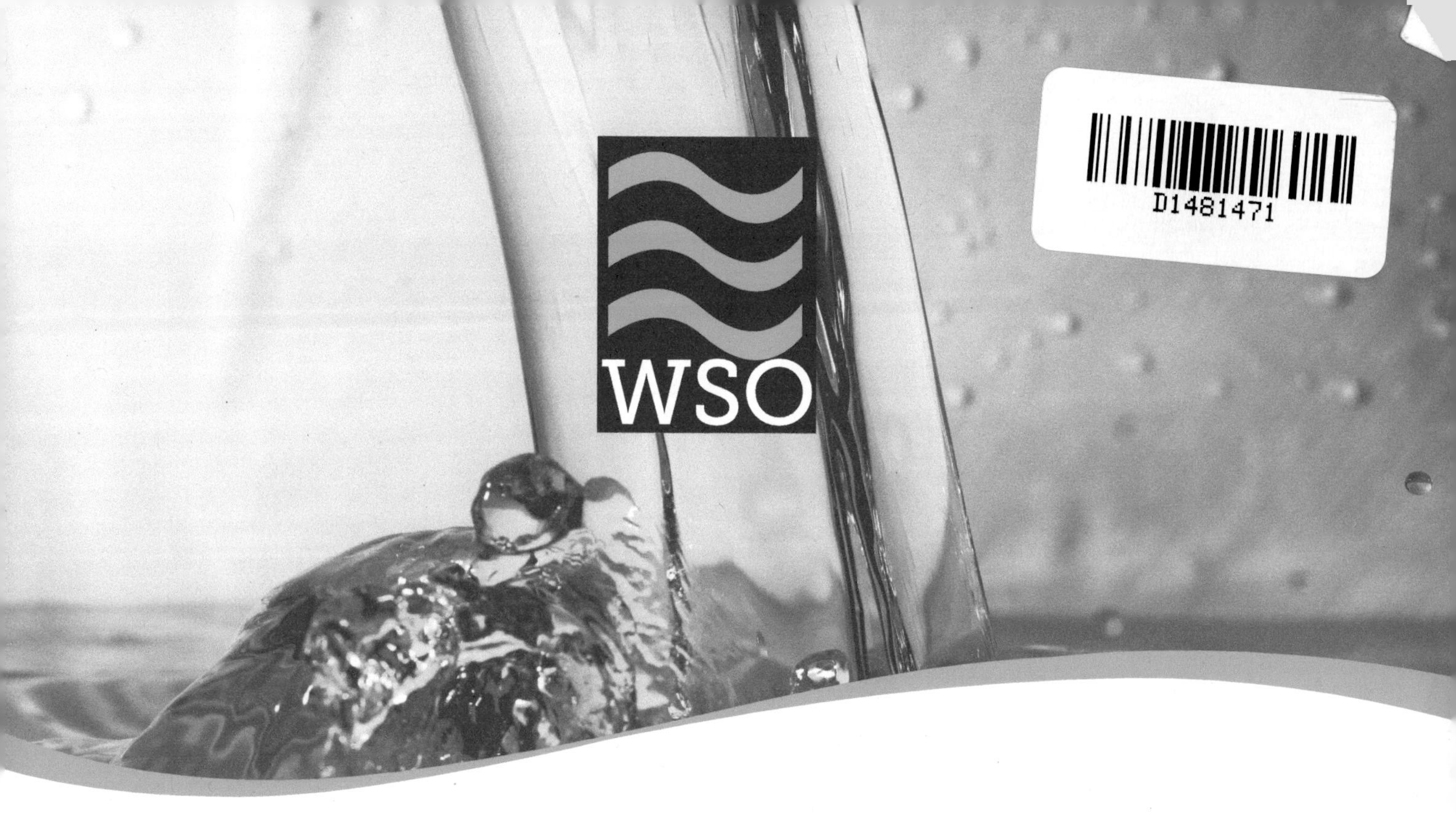

WSO

Water Distribution

Grades 3 & 4

Printed in the United States of America.

Disclaimer
Many of the photographs and illustrative drawings that appear in this book have been furnished through the courtesy of various product distributors and manufacturers. Any mention of trade names, commercial products, or services does not constitute endorsement or recommendation for use by the American Water Works Association or the US Environmental Protection Agency. In no event will AWWA be liable for direct, indirect, special, incidental, or consequential damages arising out of the use of information presented in this book. In particular, AWWA will not be responsible for any costs, including, but not limited to, those incurred as a result of lost revenue. In no event shall AWWA's liability exceed the amount paid for the purchase of this book.

Library of Congress Cataloging-in-Publication Data
CIP data have been applied for.

ISBN: 9781625761279

0002000102720233381

6666 West Quincy Avenue
Denver, CO 80235-3098
303.794.7711

Contents

Foreword

This book is part of the *Water System Operations* (WSO) series. This water operator education series was designed by the American Water Works Association (AWWA) to address core test content on certification exams by operator certification type (treatment or distribution) and certification grade level.

The current books in the series are:

WSO Water Treatment, Grade 1

WSO Water Treatment, Grade 2

WSO Water Treatment, Grades 3 & 4

WSO Water Distribution, Grades 1 & 2

WSO Water Distribution, Grades 3 & 4

Acknowledgments

The WSO series was developed by AWWA with the help of a volunteer steering committee of subject matter experts.

We would like to extend our thanks to the following individuals for their invaluable help and expertise:

- William Lauer, Project Manager, QualQuest, LLC (Lakewood, CO)
- Zac Bertz, Joint Water Commission (Hillsboro, OR)
- Mary Howell, Backflow Management, Inc. (Portland, OR)
- Bob Hoyt, City of Worcester Department of Public Works and Parks (Worcester, MA)
- Ted Kenney, New England Water Works Association (Holliston, MA)
- Darin LaFalam, City of Worcester Department of Public Works and Parks (Worcester, MA)
- Kenneth C. Morgan, KCM Consulting Services, LLC (Phoenix, AZ)
- Ray Olson, Distribution System Resources (Littleton, CO)
- Paul Riendeau, New England Water Works Association (Holliston, MA)
- Ben Wright, City of Cayce Water and Sewer Department (Cayce, SC)

Additionally, we would like to thank Barbara Martin, AWWA/Partnership for Safe Water (Denver, CO), for her technical assistance and review.

How to Use This Book

Thank you for purchasing this volume in the Water System Operations series. AWWA's WSO series conforms to the latest Association of Boards of Certification (ABC) Need-to-Know criteria. ABC administers water operator certification testing for most of North America. Some states and provinces utilize different certification testing authorities, but this book covers all of the fundamentals every water operator needs to effectively pass their certification test and do their job effectively.

To help you advance in your career as a water operator, the WSO series is divided by subject areas (water treatment and water distribution) and certification grades (1, 2, 3, and 4, following the most common practice among states and provinces). Reference material, including basic science and mathematics concepts and information on water sources and quality, is included throughout all volumes in the series.

New features have been added to help you get the most from this book:

Key words Important terms are highlighted in blue when they are first introduced, with a corresponding definition box in the margin of the page. You can also look up key words in the glossary at the end of the book.

The first pretreatment provided in most surface water treatment systems is screening. Coarse screens located on an intake structure are usually called *trash racks* or *debris racks*. Their function is to prevent clogging of the intake by removing sticks, logs, and other large debris in a river, lake, or reservoir. Finer screens may then be used at the point where the water enters the treatment system to remove smaller debris that has passed the trash racks.

The two basic types of screens used by water systems are bar screens and wire-mesh screens. Both types are available in models that are manually cleaned or automatically cleaned by mechanical equipment.

screening
A pretreatment method that uses coarse screens to remove large debris from the water to prevent clogging of pipes or channels to the treatment plant.

Bar Screens

Bar screens are made of straight steel bars, welded at both ends to two horizontal steel members. The screens are usually ranked by the open distance between bars as follows:

- Fine: spacing of 1/16 to ½ in. (1.5–13 mm)
- Medium: spacing of ½ to 1 in. (13–25 mm)
- Coarse: spacing of 1¼ to 4 in. (32–100 mm)

bar screens
A series of straight steel bars, welded at their ends to horizontal steel beams, forming a grid. Bar screens are placed on intakes or in waterways to remove large debris.

Video clips A number of subjects in this book are linked to helpful video clips available on the AWWA WSO website (www.awwa.org/wsovideoclips). The videos present visual, hands-on information to supplement the descriptions in the text.

WATCH THE VIDEO
Sedimentation and Clarifiers (www.awwa.org/wsovideoclips)

End-of-chapter questions To test your understanding of the core concepts and applications in the book, questions are provided at the end of each chapter. The answers are provided at the end of the book, with the steps worked-out for math problems.

End-of-book questions In the higher-level WSO books, questions are included at the end of the book to test your understanding of topics covered in previous grades. Keep in mind that certification tests will cover material from both lower-level and higher-level books.

Visit our website at www.awwa.org/wso to check out the additional books in the series and to access the free online resources associated with this book.

Chapter 1
USEPA Water Regulations

Regulations that govern US water supply and treatment are developed by the US Environmental Protection Agency (USEPA) under the Safe Drinking Water Act (SDWA). Most states administer USEPA regulations after adopting regulations that are no less stringent than federal rules; and in some cases, states have adopted stricter regulations or have developed regulations for additional contaminants not regulated by USEPA.

This chapter discusses current and anticipated USEPA regulations and the challenges that operators face in their efforts to comply with the regulations. Water system operators should consult their local and state regulatory agencies to verify applicable regulations that may be different than the federal regulations listed in this chapter. The chapter concludes with a discussion of selected contaminants that are commonly found in water, their significance, and the methods for their removal.

Types of Water Systems

The SDWA defines a public water system (PWS) as a supply of piped water for human consumption that has at least 15 service connections, or serves 25 or more persons 60 or more days each year. By that definition, private homes, groups of homes with a single water source but having fewer than 25 residents, and summer camps with their own water source that operate less than 60 days per year are not PWSs. They may, however, be subject to state or local regulations. Such systems may also be subject to state and local well construction and water quality requirements.

PWSs are classified into three categories based on the type of customers served:

- *Community PWS:* a system whose customers are full-time residents
- *Nontransient noncommunity PWS:* an entity having its own water supply, serving an average of at least 25 persons who do not live at the location but who use the water for more than 6 months per year
- *Transient noncommunity PWS*: an establishment having its own water system, where an average of at least 25 people per day visit and use the water occasionally or for only short periods of time

These classifications are based on the differences in exposure to contaminants experienced by persons using the water. Most chemical contaminants are believed to potentially cause adverse health effects from long-term exposure. Short-term

US Environmental Protection Agency (USEPA)
A US government agency responsible for implementing federal laws designed to protect the environment. Congress has delegated implementation of the Safe Drinking Water Act to the USEPA.

exposure to low-level chemical contamination may not carry the same risk as long-term exposure.

Therefore, the monitoring requirements for both community and noncommunity water systems apply to all contaminants that are considered a health threat. The transient and nontransient noncommunity systems must monitor only for nitrite and nitrate, as well as biological contamination (those sources that pose an immediate threat from brief exposure). The remaining community systems, about 52,000 in the United States, have more stringent and frequent monitoring requirements.

Before examining the specific regulations that govern contaminants, the operator needs to know the difference between the two concepts used in the contaminant monitoring process: the **maximum contaminant level goal (MCLG)** and the **maximum contaminant level (MCL)**:

- The MCLG is set for most substances at a level where there are no known, or anticipated, health effects. For those substances that are suspected carcinogens, the MCLG is set at zero.
- The MCL is set as close as feasible to the MCLG for substances regulated under the SDWA. The MCL is a level that is reasonably and economically achievable. This is the enforceable regulated level. Water systems that exceed an MCL must take steps to install treatment to reduce the contaminant concentration to below the MCL. Where USEPA has found it impractical to set an MCL, a treatment technique has been established instead of an MCL.

With these concepts in mind, the various regulations can be examined. This discussion is not meant to be all-inclusive. Because the regulatory process is an ever-evolving one, the reader is cautioned that some of the stated facts presented in this discussion may have changed since the writing of this chapter. For up-to-date information, it is best to contact the local office of the regulatory authority in the district or state where the utility operates.

Table 1-1 lists the status of USEPA primary drinking water standards at the time this book was prepared. These standards are provided for illustration only and are not intended to be used for regulatory purposes (see the official USEPA regulatory information on the agency website).

Operations personnel are expected to know the regulatory limits for compounds encountered in their water supply. However, the number and variety of regulated substances make it unlikely that operators would know all of the regulatory limits. Operators must rely on current references for the most accurate information. These are available from the regulatory agency responsible for the location of the treatment plant.

maximum contaminant level goal (MCLG)
Nonenforceable health-based goals published along with the promulgation of an MCL. Originally called *recommended maximum contaminant levels (RMCLs).*

maximum contaminant level (MCL)
The maximum permissible level of a contaminant in water as specified in the regulations of the Safe Drinking Water Act.

Disinfection By-product and Microbial Regulations

Drinking water treatment, including use of chemical disinfectants such as chlorine, ozone, and chlorine dioxide, has been an important step in protecting drinking water consumers from exposure to harmful microbial contaminants. However, these chemical disinfectants can also react with organic and inorganic substances in the water to produce by-products that may be harmful to

Table 1-1 List of contaminants and their MCLs

Contaminant	MCLG,* mg/L†	MCL or TT, mg/L	Potential Health Effects From Ingestion of Water	Sources of Contaminant in Drinking Water
Microorganisms				
Cryptosporidium	Zero	TT‡	Gastrointestinal illness (e.g., diarrhea, vomiting, cramps), Cryptosporidiosis	Human and animal fecal waste
Giardia lamblia	Zero	TT	Gastrointestinal illness (e.g., diarrhea, vomiting, cramps), Giardiasis	Human and animal fecal waste
Heterotrophic plate count (HPC)	N/A	TT	HPC has no health effects; it is an analytic method used to measure the variety of bacteria that are common in water. The lower the concentration of bacteria in drinking water, the better maintained the water system is.	HPC measures a range of bacteria that are naturally present in the environment
Legionella	Zero	TT	Legionnaire's disease, a type of pneumonia	Found naturally in water; multiplies in heating systems
Total coliforms (including fecal coliform and *Escherichia coli* [*E. coli*])	Zero	5.0%**	Not a health threat in itself; it is used to indicate whether other potentially harmful bacteria may be present.††	Coliforms are naturally present in the environment as well as feces; fecal coliforms and *E. coli* only come from human and animal fecal waste.
Turbidity	N/A	TT	Turbidity is a measure of the cloudiness of water. It is used to indicate water quality and filtration effectiveness (e.g., whether disease-causing organisms are present). Higher turbidity levels are often associated with higher levels of disease-causing microorganisms such as viruses, parasites, and some bacteria. These organisms can cause symptoms such as nausea, cramps, diarrhea, and associated headaches.	Soil runoff
Viruses (enteric)	Zero	TT	Gastrointestinal illness (e.g., diarrhea, vomiting, cramps)	Human and animal fecal waste
Disinfection By-products				
Bromate	Zero	0.010	Increased risk of cancer	By-product of drinking water disinfection with ozone
Chlorite	0.8	1.0	Anemia; nervous system effects in infants and young children	By-product of drinking water disinfection chlorine dioxide
Haloacetic acids (HAA5)	N/A‡‡	0.060	Increased risk of cancer	By-product of drinking water disinfection

(continued)

Table 1-1 List of contaminants and their MCLs (continued)

Contaminant	MCLG,* mg/L†	MCL or TT, mg/L	Potential Health Effects From Ingestion of Water	Sources of Contaminant in Drinking Water
Total trihalomethanes (TTHMs)	None***	0.080	Liver, kidney, or central nervous system problems; increased risk of cancer	By-product of drinking water disinfection
Disinfectants				
Chloramines (as Cl_2)	MRDLG=4	MRDL=4.0	Eye/nose irritation; stomach discomfort; anemia	Water additive used to control microbes
Chlorine (as Cl_2)	MRDLG=4	MRDL=4.0	Eye/nose irritation; stomach discomfort	Water additive used to control microbes
Chlorine dioxide (as ClO_2)	MRDLG=0.8	MRDL=0.8	Anemia; nervous system effects in infants and young children	Water additive used to control microbes
Inorganic Chemicals				
Antimony	0.006	0.006	Increase in blood cholesterol; decrease in blood sugar	Discharge from petroleum refineries; fire retardants; ceramics; electronics; new lead-free solder
Arsenic	0	0.010 as of 1/23/06	Skin damage or problems with circulatory systems; possible increased risk of contracting cancer	Erosion of natural deposits; runoff from orchards, runoff from glass and electronics production wastes
Asbestos (fiber >10 micrometers)	7 million fibers per liter (MFL)	7 MFL	Increased risk of developing benign intestinal polyps	Decay of asbestos cement in water mains; erosion of natural deposits
Barium	2	2	Increase in blood pressure	Discharge of drilling wastes; discharge from metal refineries; erosion of natural deposits
Beryllium	0.004	0.004	Intestinal lesions	Discharge from metal refineries and coal-burning factories; discharge from electrical, aerospace, and defense industries
Cadmium	0.005	0.005	Kidney damage	Corrosion of galvanized pipes; erosion of natural deposits; discharge from metal refineries; runoff from waste batteries and paints
Chromium (total)	0.1	0.1	Allergic dermatitis	Discharge from steel and pulp mills; erosion of natural deposits
Copper	1.3	TT†††; action level=1.3	Short-term exposure: gastrointestinal distress; long-term exposure: liver or kidney damage; people with Wilson's disease should consult their personal doctor if the amount of copper in their water exceeds the action level.	Corrosion of household plumbing systems; erosion of natural deposits

Table 1-1 List of contaminants and their MCLs (continued)

Contaminant	MCLG,* mg/L†	MCL or TT, mg/L	Potential Health Effects From Ingestion of Water	Sources of Contaminant in Drinking Water
Cyanide (as free cyanide)	0.2	0.2	Nerve damage or thyroid problems	Discharge from steel/metal factories; discharge from plastics and fertilizer factories
Fluoride	4.0	4.0	Bone disease (pain and tenderness of the bones); children may get mottled teeth	Water additive that promotes strong teeth; erosion of natural deposits; discharge from fertilizer and aluminum factories
Lead	Zero	TT; action level=0.015	Infants and children: Delays in physical or mental development; children could show slight deficits in attention span and learning abilities Adults: Kidney problems; high blood pressure	Corrosion of household plumbing systems; erosion of natural deposits
Mercury (inorganic)	0.002	0.002	Kidney damage	Erosion of natural deposits; discharge from refineries and factories; runoff from landfills and croplands
Nitrate (measured as nitrogen)	10	10	Infants below the age of 6 months who drink water containing nitrate in excess of the MCL could become seriously ill and, if untreated, may die. Symptoms include shortness of breath and blue-baby syndrome.	Runoff from fertilizer use; leaching from septic tanks, sewage; erosion of natural deposits
Nitrite (measured as nitrogen)	1	1	Infants below the age of 6 months who drink water containing nitrite in excess of the MCL could become seriously ill and, if untreated, may die. Symptoms include shortness of breath and blue-baby syndrome.	Runoff from fertilizer use; leaching from septic tanks, sewage; erosion of natural deposits
Selenium	0.05	0.05	Hair or fingernail loss; numbness in fingers or toes; circulatory problems	Discharge from petroleum refineries; erosion of natural deposits; discharge from mines
Thallium	0.0005	0.002	Hair loss; changes in blood; kidney, intestine, or liver problems	Leaching from ore-processing sites; discharge from electronics, glass, and drug factories
Organic Chemicals—Synthetic Organic Chemicals (SOCs)				
2,4,5-TP (Silvex)	0.05	0.05	Liver problems	Residue of banned herbicide
2,4-D	0.07	0.07	Kidney, liver, or adrenal gland problems	Runoff from herbicide used on row crops

(continued)

Table 1-1 List of contaminants and their MCLs (continued)

Contaminant	MCLG,* mg/L†	MCL or TT, mg/L	Potential Health Effects From Ingestion of Water	Sources of Contaminant in Drinking Water
Acrylamide	Zero	TT‡‡‡	Nervous system or blood problems; increased risk of cancer	Added to water during sewage/wastewater treatment
Alachlor	Zero	0.002	Eye, liver, kidney, or spleen problems; anemia; increased risk of cancer	Runoff from herbicide used on row crops
Atrazine	0.003	0.003	Cardiovascular system or reproductive problems	Runoff from herbicide used on row crops
Benzo(a)pyrene (PAHs)	Zero	0.0002	Reproductive difficulties; increased risk of cancer	Leaching from linings of water storage tanks and distribution lines
Carbofuran	0.04	0.04	Problems with blood, nervous system, or reproductive system	Leaching of soil fumigant used on rice and alfalfa
Chlordane	Zero	0.002	Liver or nervous system problems; increased risk of cancer	Residue of banned termiticide
Dalapon	0.2	0.2	Minor kidney changes	Runoff from herbicide used on rights-of-way
Di(2-ethylhexyl) adipate	0.4	0.4	Weight loss; liver problems; possible reproductive difficulties	Discharge from chemical factories
Di(2-ethylhexyl) phthalate (DEHP)	Zero	0.006	Reproductive difficulties; liver problems; increased risk of cancer	Discharge from rubber and chemical
1,2-Dibromo-3-chloropropane (DBCP)	Zero	0.0002	Reproductive difficulties; increased risk of cancer	Runoff/leaching from soil fumigant used on soybeans, cotton, pineapples, and orchards
Dinoseb	0.007	0.007	Reproductive difficulties	Runoff from herbicide used on soybeans and vegetables
Diquat	0.02	0.02	Cataracts	Runoff from herbicide use
Dioxin (2,3,7,8-TCDD)	Zero	0.00000003	Reproductive difficulties; increased risk of cancer	Emissions from waste incineration and other combustion; discharge from chemical factories
Endothall	0.1	0.1	Stomach and intestinal problems	Runoff from herbicide use
Endrin	0.002	0.002	Liver problems	Residue of banned insecticide
Epichlorohydrin	Zero	TT	Increased cancer risk, and over a long period of time, stomach problems	Discharge from industrial chemical factories; an impurity of some water treatment chemicals
Ethylene dibromide	Zero	0.00005	Problems with liver, stomach, reproductive system, or kidneys; increased risk of cancer	Discharge from petroleum refineries

Table 1-1 List of contaminants and their MCLs (continued)

Contaminant	MCLG,* mg/L†	MCL or TT, mg/L	Potential Health Effects From Ingestion of Water	Sources of Contaminant in Drinking Water
Glyphosate	0.7	0.7	Kidney problems; reproductive difficulties	Runoff from herbicide use
Heptachlor	Zero	0.0004	Liver damage; increased risk of cancer	Residue of banned termiticide
Heptachlor epoxide	Zero	0.0002	Liver damage; increased risk of cancer	Breakdown of heptachlor
Hexachloro-benzene	Zero	0.001	Liver or kidney problems; reproductive difficulties; increased risk of cancer	Discharge from metal refineries and agricultural chemical factories
Hexachloro-cyclopentadiene (HEX)	0.05	0.05	Kidney or stomach problems	Discharge from chemical factories
Lindane	0.0002	0.0002	Liver or kidney problems	Runoff/leaching from insecticide used on cattle, lumber, gardens
Methoxychlor	0.04	0.04	Reproductive difficulties	Runoff/leaching from insecticide used on fruits, vegetables, alfalfa, livestock
Oxamyl (Vydate)	0.2	0.2	Slight nervous system effects	Runoff/leaching from insecticide used on apples, potatoes, and tomatoes
Pentachloro-phenol	Zero	0.001	Liver or kidney problems; increased cancer risk	Discharge from wood-preserving factories
Picloram	0.5	0.5	Liver problems	Herbicide runoff
Polychlorinated biphenyls (PCBs)	Zero	0.0005	Skin changes; thymus gland problems; immune deficiencies; reproductive or nervous system difficulties; increased risk of cancer	Runoff from landfills; discharge of waste chemicals
Simazine	0.004	0.004	Problems with blood	Herbicide runoff
Toxaphene	Zero	0.003	Kidney, liver, or thyroid problems; increased risk of cancer	Runoff/leaching from insecticide used on cotton and cattle
Organic Chemicals—Volatile Organic Chemicals (VOCs)				
Benzene	Zero	0.005	Anemia; decrease in blood platelets; increased risk of cancer	Discharge from factories; leaching from gas storage tanks and landfills
Chlorobenzene	0.1	0.1	Liver or kidney problems	Discharge from chemical and agricultural chemical factories
Carbon tetra-chloride	Zero	0.005	Liver problems; increased risk of cancer	Discharge from chemical plants and other industrial activities
o-Dichloro-benzene	0.6	0.6	Liver, kidney, or circulatory system problems	Discharge from industrial chemical factories
p-Dichloro-benzene	0.075	0.075	Anemia; liver, kidney, or spleen damage; changes in blood	Discharge from industrial chemical factories

(continued)

Table 1-1 List of contaminants and their MCLs (continued)

Contaminant	MCLG,* mg/L†	MCL or TT, mg/L	Potential Health Effects From Ingestion of Water	Sources of Contaminant in Drinking Water
1,2-Dichloro-ethane	Zero	0.005	Increased risk of cancer	Discharge from industrial chemical factories
1,1-Dichloro-ethylene	0.007	0.007	Liver problems	Discharge from industrial chemical factories
cis-1,2-Dichloro-ethylene	0.07	0.07	Liver problems	Discharge from industrial chemical factories
trans-1,2-Dichloroethylene	0.1	0.1	Liver problems	Discharge from industrial chemical factories
Dichloromethane	Zero	0.005	Liver problems; increased risk of cancer	Discharge from drug and chemical factories
1,2-Dichloro-propane	Zero	0.005	Increased risk of cancer	Discharge from industrial chemical factories
Ethylbenzene	0.7	0.7	Liver or kidney problems	Discharge from petroleum refineries
Styrene	0.1	0.1	Liver, kidney, or circulatory system problems	Discharge from rubber and plastics factories; leaching from landfills
Tetrachloro-ethylene (PCE)	Zero	0.005	Liver problems; increased risk of cancer	Discharge from factories and dry cleaners
Toluene	1	1	Nervous system, kidney, or liver problems	Discharge from petroleum factories
1,2,4-Trichloro-benzene	0.07	0.07	Changes in adrenal glands	Discharge from textile finishing factories
1,1,1-Trichloro-ethane	0.2	0.2	Liver, nervous system, or circulatory problems	Discharge from metal degreasing sites and other factories
1,1,2-Trichloro-ethane	0.003	0.005	Liver, kidney, or immune system problems	Discharge from industrial chemical factories
Trichloroethylene (TCE)	Zero	0.005	Liver problems; increased risk of cancer	Discharge from metal degreasing sites and other factories
Vinyl chloride	Zero	0.002	Increased risk of cancer	Leaching from PVC pipes; discharge from plastics factories
Xylenes (total)	10	10	Nervous system damage	Discharge from petroleum factories; discharge from chemical factories
Radionuclides				
Alpha particles	None Zero	15 picocuries per liter (pCi/L)	Increased risk of cancer	Erosion of natural deposits of certain minerals that are radioactive and may emit a form of radiation known as alpha radiation
Beta particles and photon emitters	None Zero	4 millirems per year	Increased risk of cancer	Decay of natural and synthetic deposits of certain minerals that are radioactive and may emit forms of radiation known as photons and beta radiation

Table 1-1 List of contaminants and their MCLs (continued)

Contaminant	MCLG,* mg/L†	MCL or TT, mg/L	Potential Health Effects From Ingestion of Water	Sources of Contaminant in Drinking Water
Radium 226 and radium 228 (combined)	None Zero	5 pCi/L	Increased risk of cancer	Erosion of natural deposits
Uranium	Zero	30 µg/L as of 12/8/03	Increased risk of cancer, kidney toxicity	Erosion of natural deposits

*Definitions:

Maximum contaminant level (MCL)—The highest level of a contaminant that is allowed in drinking water. MCLs are set as close to MCLGs as feasible using the best available treatment technology and taking cost into consideration. MCLs are enforceable standards.

Maximum contaminant level goal (MCLG)—The level of a contaminant in drinking water below which there is no known or expected risk to health. MCLGs allow for a margin of safety and are nonenforceable public health goals.

Maximum residual disinfectant level (MRDL)—The highest level of a disinfectant allowed in drinking water. There is convincing evidence that addition of a disinfectant is necessary for control of microbial contaminants.

Maximum residual disinfectant level goal (MRDLG)—The level of a drinking water disinfectant below which there is no known or expected risk to health. MRDLGs do not reflect the benefits of the use of disinfectants to control microbial contaminants.

Treatment Technique (TT)—A required process intended to reduce the level of a contaminant in drinking water.

†Units are in milligrams per liter (mg/L) unless otherwise noted. Milligrams per liter is equivalent to parts per million.

‡USEPA's Surface Water Treatment Rules (SWTRs) require systems using surface water or groundwater under the direct influence of surface water to (1) disinfect their water and (2) filter their water or meet criteria for avoiding filtration so that the following contaminants are controlled at the following levels:

Cryptosporidium (as of Jan. 1, 2002, for systems serving >10,000 and 1/14/05 for systems serving <10,000) 99% removal.

Giardia lamblia: 99.9% removal/inactivation.

Viruses: 99.99% removal/inactivation.

Legionella: No limit, but USEPA believes that if *Giardia* and viruses are removed/inactivated, *Legionella* will also be controlled.

Turbidity: At no time can turbidity (cloudiness of water) go above 5 nephelometric turbidity units (ntu); systems that filter must ensure that the turbidity go no higher than 1 ntu (0.5 ntu for conventional or direct filtration) in at least 95% of the daily samples in any month. As of Jan. 1, 2002, turbidity may never exceed 1 ntu and must not exceed 0.3 ntu 95% of daily samples in any month.

HPC: No more than 500 bacterial colonies per milliliter.

Long-Term 1 Enhanced Surface Water Treatment Rule (effective date: Jan. 14, 2005); surface water systems or groundwater under the direct influence of surface water (GWUDI) systems serving fewer than 10,000 people must comply with the applicable Long-Term 1 Enhanced Surface Water Treatment Rule provisions (e.g., turbidity standards, individual filter monitoring, *Cryptosporidium* removal requirements, updated watershed control requirements for unfiltered systems).

Filter Backwash Recycling: The Filter Backwash Recycling Rule requires systems that recycle to return specific recycle flows through all processes of the system's existing conventional or direct filtration system or at an alternate location approved by the state.

**More than 5.0% samples total coliform-positive in a month. (For water systems that collect fewer than 40 routine samples per month, no more than one sample can be total coliform–positive per month.) Every sample that has total coliform must be analyzed for either fecal coliforms or E. coli; if two consecutive total coliform samples are positive and one is also positive for *E. coli* fecal coliforms, the system has an acute MCL violation.

††Fecal coliform and *E. coli* are bacteria whose presence indicates that the water may be contaminated with human or animal wastes. Disease-causing microbes (pathogens) in these wastes can cause diarrhea, cramps, nausea, headaches, or other symptoms. These pathogens may pose a special health risk for infants, young children, and people with severely compromised immune systems.

‡‡Although there is no collective maximum contaminant level goal (MCLG) for this contaminant group, there are individual MCLGs for some of the individual contaminants: trihalomethanes: bromodichloromethane (zero); bromoform (zero); dibromochloromethane (0.06 mg/L). Chloroform is regulated with this group but has no MCLG. Haloacetic acids: dichloroacetic acid (zero); trichloroacetic acid (0.3 mg/L). Monochloroacetic acid, bromoacetic acid, and dibromoacetic acid are regulated with this group but have no MCLGs.

***MCLGs were not established before the 1986 amendments to the SDWA. Therefore, there is no MCLG for this contaminant.

†††Lead and copper are regulated by a treatment technique that requires systems to control the corrosiveness of their water. If more than 10% of tap water samples exceed the action level, water systems must take additional steps. For copper, the action level is 1.3 mg/L; for lead it is 0.015 mg/L.

‡‡‡Each water system must certify in writing to the state (using third-party or manufacturer's certification) that when acrylamide and epichlorohydrin are used in drinking water systems, the combination (or product) of dose and monomer level does not exceed the levels specified, as follows:

Acrylamide = 0.05% dosed at 1 mg/L (or equivalent)

Epichlorohydrin = 0.01% dosed at 20 mg/L (or equivalent)

drinking water consumers, particularly some susceptible segments of the population. Therefore, drinking water treatment using chemical disinfectants involves a delicate balancing act—i.e., adding enough disinfectant to control harmful microorganisms but not enough to produce unacceptably high levels of regulated **disinfection by-products (DBPs)**.

USEPA has enacted several regulations impacting microbial control and production of DBPs in groundwater and surface water supplies for small and large public drinking water systems. These rules are referred to collectively as the Microbial/Disinfection By-Products (M/DBP) Rules. Microbial protection for consumers of drinking water from public supplies is provided by provisions of current or pending rules listed below and discussed in more detail later in this chapter:

- Filter Backwash Recycling Rule (FBRR)
- Ground Water Rule (GWR)
- Interim Enhanced Surface Water Treatment Rule (IESWTR)
- Long-Term 1 Enhanced Surface Water Treatment Rule (LT1ESWTR)
- Long-Term 2 Enhanced Surface Water Treatment Rule (LT2ESWTR)
- Stage 1 Disinfectants and Disinfection By-products Rule (Stage 1 DBPR)
- Stage 2 Disinfectants and Disinfection By-products Rule (Stage 2 DBPR)
- Surface Water Treatment Rule (SWTR)
- Total Coliform Rule (TCR)

Provisions of the Disinfectants and Disinfection By-products Rule (DBPR) are intended to protect drinking water consumers against the unintended public health consequences associated with consumption of treated drinking water containing residual disinfectants and DBPs produced from degradation of these residual disinfectants or reaction of disinfectants with organic and inorganic DBP precursors.

More details regarding the DBPR, including the current Stage 1 DBPR and Stage 2 DBPR, are described in this chapter. Also included in the DBPR description is a brief discussion of some currently unregulated DBPs that are being heavily researched and may be the subject of future regulation. In the following discussion, the DBPR will be discussed first, followed by the microbial protection rules (SWTR, GWR, and TCR).

Disinfection By-product Rule (DBPR)

The Stage 1 DBPR and Stage 2 DBPR requirements discussed in the following sections focus first on two specific contaminants (TTHM and HAA5) and then on other aspects of these regulations dealing with control or removal of DBP precursors ("enhanced coagulation"), bromate, chlorite, and residual disinfectants.

WATCH THE VIDEO
Disinfection By-products (www.awwa.org/wsovideoclips)

disinfection by-products (DBPs)
New chemical compounds that are formed by the reaction of disinfectants with organic compounds in water. At high concentrations, many DBPs are considered a danger to human health.

Stage 1 DBPR—HAA5 and TTHM Provisions

The Stage 1 DBPR was published in 1998 and established an MCL of 0.080 mg/L for TTHM (the sum of four trihalomethanes, which are chloroform, bromodichloromethane, dibromochloromethane, and bromoform) and 0.060 mg/L for HAA5 (the sum of five specific haloacetic acids, which are mono-, di-, and trichloroacetic acids plus mono- and dibromoacetic acids). Although the MCLs for TTHM and HAA5 were officially written as 0.080 mg/L and 0.060 mg/L, respectively, the limits are commonly referred to as "80/60," or 80 μg/L and 60 μg/L. While knowing the numerical value of each MCL is important in understanding compliance with the DBPR, it is equally important to understand the methodology, in all its subtleties, used to calculate the compliance value that will be compared to this MCL.

For TTHM and HAA5, the compliance value is determined by monitoring the distribution system. Compliance monitoring locations need to be representative of the distribution system. Systems serving more than 10,000 persons who use surface water sources are required to monitor at least four locations per plant, meaning that distribution systems fed by more than one treatment plant must have at least four monitoring locations designated for each plant entry point.

The compliance monitoring location for systems with only one monitoring point must be representative of maximum residence time in the distribution system. A minimum of one out of every four compliance monitoring locations for systems with more locations must also be representative of maximum residence time. The other locations must be far enough away from the plant entry points to be representative of average residence time in the distribution system.

Unlike acute toxicity risks, for which the exposure could be a single glass of water, cancer risks like those believed to be linked to TTHM and HAA5 involve longer periods of exposure (daily glasses of water spanning decades). For chronic exposures such as these, exposure to an excessively high concentration of a given cancer-causing agent will not necessarily result in the consumer getting cancer from this source. Conversely, a consumer exposed to a lower concentration every day for a lifetime could be more likely to develop cancer. Therefore, regulation of DBPs to reduce cancer risks is *not* based on limiting exposure to a single incident (i.e., not a "single hit"), but rather is aimed at reducing the repeated exposure over time. In other words, DBP exposure needs to be evaluated on an average basis over time.

Under the Stage 1 DBPR, the compliance value for TTHM and HAA5 is determined by calculating a **running annual average (RAA)** during the previous 12 months for each DBP for all monitoring locations at each plant. Most systems are required to monitor quarterly (i.e., 4 times per year), although small groundwater systems (<10,000 persons) may be allowed to sample once a year. Typically, the RAA is based on 4 monitoring locations sampled quarterly, meaning RAA will be the average of 16 monitoring results each for HAA5 and TTHM.

Table 1-2 illustrates one facility's calculations of RAA for HAA5 that were used for Stage 1 compliance (this table also shows calculation of values for Stage 2 DBPR, which will be discussed later). It is important to reemphasize that compliance is based solely on the RAA, not on a single quarterly result at any one monitoring location. Consequently, it is *not* correct to refer to a single quarterly monitoring result above 60 μg/L for HAA5 or above 80 μg/L for TTHM as being above the MCL. Therefore, even though several individual monitoring values

Table 1-2 Example RAA and LRAA calculations for Stage 1 DBPR and Stage 2 DBPR

		Sampling Location, μg/L			
Year	**Quarter**	**A**	**B**	**C**	**D**
1	3rd	52	68	63	66
1	4th	35	42	38	41
2	1st	47	49	42	43
2	2nd	18	42	45	37
LRAA		38	50	47	47
Maximum LRAA			50		
RAA			45		

running annual average (RAA)
The average of four quarterly samples at each monitoring location to ensure compliance with the Stage 2 DBPR.

in Table 1-2 are greater than 60 µg/L, the facility is in compliance with the HAA5 MCL because the RAA is 45 µg/L for HAA5.

Utility personnel should be consistent and rigorous in their use of terminology when dealing with the general public or with state and local health officials, and should ensure that all people participating in these discussions are consistent in applying the MCL only to RAA values and do not make the common mistake of referring to a single quarterly monitoring value as being "above the MCL."

Stage 2 DBPR—HAA5 and TTHM Provisions

The Stage 2 DBPR, published in 2006, is now in effect. This rule tightened requirements for DBPs, but compliance is not achieved by modifying the numerical value of the MCLs or by requiring monitoring of new constituents. Instead, the rule makes compliance more challenging than under the Stage 1 DBPR by (1) changing the way the compliance value is calculated and (2) changing the compliance monitoring locations to sites representative of the greatest potential for THM and HAA formation. These changes were made to ensure uniform compliance with the DBP standards across all areas of the distribution system; that is, compliance is required at each sampling location.

The compliance value in the Stage 2 DBPR is called the *locational running annual average* (LRAA), and it is calculated by separately averaging the four quarterly samples at each monitoring location. Compliance is based on the maximum LRAA value (see Table 1-2). Furthermore, the Stage 2 DBPR included several interim steps that led to the replacement of many existing Stage 1 DBPR monitoring locations with new locations representative of the greatest potential for consumer exposure to high levels of TTHM and HAA5.

The Stage 2 DBPR required that facilities maintain compliance with the Stage 1 DBPR using the existing monitoring locations during the first three years after the final version of the Stage 2 DBPR was published. In the time period between the third and sixth year after the Stage 2 DBPR was published, compliance continues to be based on maintaining 80/60 (TTHM and HAA5) or lower for RAA; it also includes a requirement for maximum LRAA at existing Stage 1 monitoring locations. The long-term goal of the Stage 2 DBPR was to identify locations within the distribution system with the greatest potential for either TTHM or HAA5 formations and then base compliance on the LRAA at or below 80/60 for each of these locations. Many of these locations were identified during the initial distribution system evaluation (IDSE).

The IDSE included monitoring, modeling, and/or other evaluations of drinking water distribution systems to identify locations representative of the greatest potential for consumer exposure to high levels of TTHM and HAA5. The goal of the IDSE was to evaluate a number of potential monitoring locations to justify selection of monitoring locations for long-term compliance (i.e., Stage 2B) with the Stage 2 DBPR.

One item to note regarding the Stage 2 DBPR as it applies to TTHM and HAA5 is that the goal was to find the locations in the distribution system where average annual levels of these DBPs are highest. TTHM formation increases as contact time with free or combined chlorine increases, although formation in the presence of combined chlorine is limited. Therefore, establishing points in the distribution system with the highest potential for TTHM formation is related to knowing the points with maximum water age. Utilities that have not performed a tracer study in the distribution system to determine water age should consider doing so.

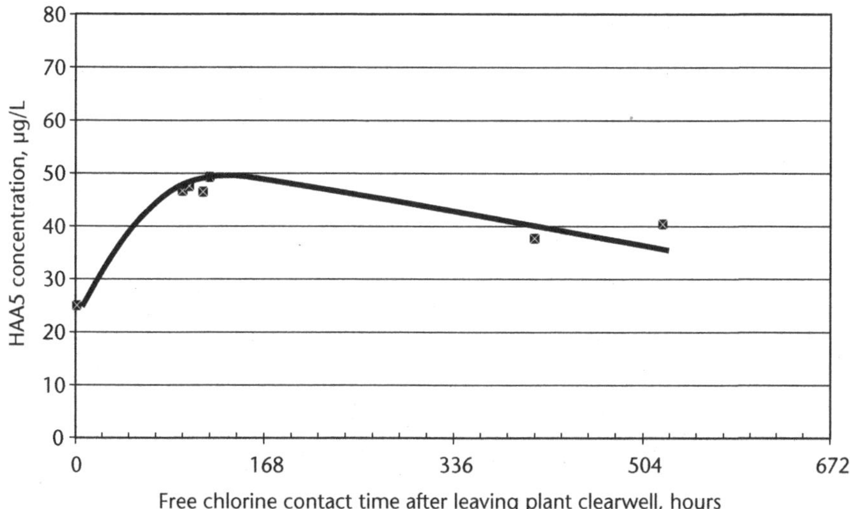

Figure 1-1 Formation and decay of HAA5 in a distribution system (time estimated by fluoride tracer test—T_{100})

By contrast, peak locations for HAA5 are more complicated because microorganisms in biofilm attached to distribution system pipe surfaces can biodegrade HAA5. Consequently, increasing formation of HAA5 over time is offset by biodegradation, eventually reaching a point where HAA5 levels decrease over time, even to the point where they drop to zero. Figure 1-1 shows a gradual reduction in HAA5 formation over time in a distribution system, followed by an eventual decrease of HAA5 as water age increases (water age measured in tracer test). In chloramination systems, HAA5 formation is limited. In fact, ammonium chloride is added as a quenching agent in HAA5 compliance samples in order to halt HAA5 formation prior to analysis (*Standard Methods for the Examination of Water and Wastewater*, latest edition). Therefore, little additional HAA5 formation occurs after chloramination to offset HAA5 biodegradation occurring in the distribution system.

Enhanced Coagulation Requirement of the Stage 1 DBPR and Stage 2 DBPR

The enhanced coagulation requirement has been developed to promote optimization of coagulation processes in conventional surface water treatment systems as required to improve removal of organic DBP precursors. The focus of the SWTR is separate from that of the enhanced coagulation requirement, with the former directed toward optimizing particle removal and the latter toward optimizing removal of natural organic matter (DBP precursors). Both promote efforts by water utilities to properly control and optimize coagulation processes and reduce DBP formation.

Under the enhanced coagulation requirements, treatment plants must remove specific percentages of total organic carbon (TOC) based on their source water TOC and alkalinity levels. Facilities must meet the enhanced coagulation requirements unless they meet any of the following exemptions (USEPA Stage 1 DBPR Guidance):

1. The PWS's source water TOC level is <2.0 mg/L, calculated quarterly as an RAA.
2. The PWS's treated water TOC level is <2.0 mg/L, calculated quarterly as an RAA.

3. The PWS's source water TOC level is <4.0 mg/L, calculated quarterly as an RAA; the source-water alkalinity is >60 mg/L (as calcium carbonate [$CaCO_3$]), calculated quarterly as a running annual average; and either the TTHM and HAA5 running annual averages are no greater than 0.040 mg/L and 0.030 mg/L, respectively, or the PWS has made a clear and irrevocable financial commitment to use technologies that will limit the levels of TTHMs and HAA5 to no more than 0.040 mg/L and 0.030 mg/L, respectively.
4. The PWS's TTHM and HAA5 RAAs are no greater than 0.040 mg/L and 0.030 mg/L, respectively, and the PWS uses only chlorine for primary disinfection and maintenance of a residual in the distribution system.
5. The PWS's source water specific ultraviolet absorption at 254 nm (SUVA), prior to any treatment and measured monthly, is ≤2.0 L/mg-m, calculated quarterly as an RAA.
6. The PWS's finished-water SUVA, measured monthly, is ≤2.0 L/mg-m, calculated quarterly as an RAA.

Additionally, alternative compliance criteria for softening systems include the following:

7. Softening that results in lowering the treated water alkalinity to <60 mg/L (as $CaCO_3$), measured monthly and calculated quarterly as an RAA.
8. Softening that results in removing at least 10 mg/L of magnesium hardness (as $CaCO_3$), measured monthly and calculated quarterly as an RAA.

Utilities that cannot meet these avoidance criteria should know their enhanced coagulation endpoint, identified as the coagulant dosage and/or pH value that, when achieved, no longer produces significant TOC reduction. Specifically, when the source water TOC is not reduced by at least 0.3 mg/L with an incremental dosage increase of 10 mg/L alum (or equivalent ferric salt) and the pH value of the source reaches a value listed in Table 1-3, the enhanced coagulation endpoint has been reached.

If a utility is not exempt, a number of steps have to be evaluated relating to TOC removal, alkalinity of source water, range of source water TOC, required TOC removal for given source water characteristics, and several other factors.

Bromate

The bromate MCL from the Stage 1 DBPR remained at 0.010 mg/L for the Stage 2 DBPR. Bromate can be present in systems using ozone that have bromide present at the ozone application point. Bromate is also potentially formed during manufacture and storage of sodium hypochlorite. Consequently, systems using

Table 1-3 Target pH values for coagulation when TOC removal rates are not sufficient

Alkalinity (mg/L $CaCO_3$)	pH Value
0–60	5.5
>60–120	6.3
>120–240	7.0
>240	7.5

ozone for oxidation or disinfection are required to monitor once a month at a distribution system entry point for bromate, but systems without ozone are not required to perform this monitoring. Systems that use ozone and that also add sodium hypochlorite will need to closely monitor the quality of these sodium hypochlorite products for bromate content.

Chlorite

Similar to bromate, chlorite monitoring is required only for systems using chlorine dioxide as an oxidant or disinfectant. Chlorite is a degradation product of chlorine dioxide. Chlorate is also a degradation product of chlorine dioxide but is not currently regulated. Chlorite and chlorate are potential degradation products of sodium hypochlorite, but systems using sodium hypochlorite are not required to monitor for chlorite unless they also use chlorine dioxide.

Monitoring for chlorite is more complicated than for bromate because chlorine dioxide will degrade and chlorite formation will increase over time. Therefore, chlorite monitoring requirements include daily monitoring at the distribution system entry point and monthly samples at three locations in the distribution system (first customer, average residence time, maximum residence time). Unlike the health risks for bromate, TTHM, and HAA5, the risk for chlorite requires compliance based on the average of the three chlorite-monitoring locations each month. The Stage 1 DBPR MCL for chlorite is 1.0 mg/L.

Residual Disinfectants

The maximum residual disinfectant level (MRDL) for combined or total chlorine is 4.0 mg/L as Cl_2. These values are based on the same data used to monitor minimum free and combined chlorine levels in the distribution system as required by the SWTR, using the same monitoring locations used for the TCR. Chlorine dioxide residual also has an MRDL of 0.8 mg/L as ClO_2, based on daily samples at the treatment plant.

Surface Water Treatment Rule (SWTR)

IESWTR and LT1ESWTR

The goal of the IESWTR is to limit human exposure to harmful organisms, including *Cryptosporidium*, by promoting achievement of particle and turbidity removal targets for surface water treatment systems. Among the IESWTR requirements that apply to surface water treatment plants are the following:

- Combined filter effluent turbidity must be ≤0.3 ntu for 95 percent of samples collected each month, including none with >1 ntu. Compliance is based on combined filter effluent samples collected at 4-hour intervals during the entire month).
- The utility must monitor each individual filter for turbidity at 15-minute intervals and must report results, including a filter profile (graphical representation of filter performance), if either of the following two conditions are met: (1) turbidity in any filter for two consecutive 15-minute intervals exceeds 1 ntu or (2) turbidity during the first 4 hours of a given filter run exceeds 0.5 ntu for two consecutive 15-minute samples. Results must be reported within 10 days of the end of the month.
- Any newly constructed finished water reservoirs must include covers to keep out dust, debris, birds, etc.

- The utility must complete a sanitary survey every 3 years. Existing surveys conducted after December 1995 can be used if they meet minimum requirements. Variances can be granted to decrease frequency to 5 years. The IESWTR explicitly requires that sanitary surveys include efforts to evaluate and control *Cryptosporidium*, in addition to other target organisms.
- Systems where the average of quarterly TTHM or HAA5 values exceeds 64 and 48 µg/L, respectively, need to complete disinfection profiling and benchmarking. *Profiling* involves determination of $C \times T$ values (concentration of disinfectant × contact time) for each segment of the treatment plant (see later discussion). *Benchmarking* involves determining the lowest monthly average during 12-month monitoring of *Giardia* and virus inactivation. This procedure is required for any systems that are considering a major change to their disinfection practice. Consultation with the primacy agency is also required before any disinfection change.
- Turbidity monitoring records must be maintained for a minimum of 3 years.

Facilities in compliance with these requirements, chiefly the turbidity monitoring provisions, are designated by the IESWTR to have provided 2-log virus removal, 2.5-log *Giardia* removal, and 2-log *Cryptosporidium* removal. Literature and other information cited in the IESWTR final rule indicate that these credits are conservative, and most facilities meeting these requirements are probably achieving far greater levels of virus, *Giardia*, and *Cryptosporidium* removal than the minimum credits previously cited. The level of *Cryptosporidium* protection cited is sufficient to meet all requirements of the IESWTR, but the rule requires a total of 3.0 credits for *Giardia* and 4.0 credits for viruses. The additional credits (0.5-log for *Giardia* and 2-log for viruses) are required to be achieved by disinfection with free chlorine, chloramines, ozone, or chlorine dioxide by meeting $C \times T$ requirements described later in this chapter.

Provisions of the IESWTR apply to large systems (>10,000 persons) using surface water sources. However, similar provisions are applied to smaller surface water systems (<10,000 persons), as outlined in the LT1ESWTR. The objectives of the LT1ESWTR and IESWTR are identical, though some of the compliance deadlines and other regulatory provisions are slightly different based on greater financial and personnel resources for larger systems.

Sanitary Surveys

Sanitary surveys are a requirement of the Interim Enhanced Surface Water Treatment Rule (IESWTR). A sanitary survey is "an onsite review of the water source, facilities, equipment, operation, and maintenance of the public water system for the purpose of evaluating the adequacy of such source, facilities, equipment, operation, and maintenance for producing and distributing safe drinking water." Surveys are usually performed by the state primacy agency and are required of all surface water systems and groundwater systems under the direct influence of surface water.

These surveys are typically divided into eight main sections, although some state primacy groups may have more:

1. Water sources
2. Water treatment process
3. Water supply pumps and pumping facilities
4. Storage facilities
5. Distribution systems

6. Monitoring, reporting, and data verification
7. Water system management and operations
8. Operator compliance with state requirements

Sanitary surveys are required on a periodic basis usually every 3 years. Surveys may be comprehensive or focused according to the regulatory agency requirements.

C × T Requirements

Every water system that uses surface water as a source must meet treatment technique requirements for the removal and/or inactivation of *Giardia*, viruses, *Legionella*, and other bacteria. Because these pathogens are not easily identified in the laboratory on a routine basis, USEPA has set quality goals in lieu of MCLs in this instance. Meeting SWTR treatment technique goals demonstrates all or part of the required microbial protection, as previously noted, but additional protection is required through the use of approved disinfection treatment chemicals. The effectiveness of disinfection depends on the type of disinfectant chemical used, the residual concentration, the amount of time the disinfectant is in contact with the water, the water temperature, and, when chlorine is used, the pH of the water.

According to USEPA, a combination of the residual concentration, C, of a disinfectant (in milligrams per liter) multiplied by the contact time, *T* (in minutes), can be used as a measure of the disinfectant's effectiveness in killing or inactivating microorganisms. For water plant operators, this means that high residuals held for a short amount of time or low residuals held for a long period of time will produce similar results. Water plants are required to provide this computation daily, and the measure must always be higher than the required minimum value.

Long-Term 2 Enhanced SWTR (LT2ESWTR)

The LT2ESWTR supplements the SWTR requirements contained in the IESWTR for large surface water systems (>10,000 persons) and the LT1ESWTR for small systems (<10,000 persons). Details of the rules can be reviewed in the *Federal Register* or at the USEPA website (http://water.epa.gov/drink/index.cfm). One of the key elements of the LT2ESWTR was the use of *Cryptosporidium* monitoring results to classify surface water sources into one of four USEPA-defined risk levels called "bins." Facilities in the lowest bin (bin 1) are required to maintain compliance with the current IESWTR. Facilities in higher bins (bins 2 to 4) are required to either (1) provide additional *Cryptosporidium* protection from new facilities or programs not currently in use at a facility or (2) demonstrate greater *Cryptosporidium* protection capabilities of existing facilities and programs using a group of USEPA-approved treatment technologies, watershed programs, and demonstration studies, referred to collectively as the *Microbial Toolbox*.

Implementation of the LT2ESWTR was phased over many years according to system size. Four separate size categories were established (schedules 1–4, with 4 being the smallest at <10,000 population) for implementing the rule. The rule for schedule-4 systems allows filtered supplies to perform initial monitoring for fecal coliform to determine if *Cryptosporidium* monitoring is required.

Filter Backwash Recycle Rule (FBRR)

The FBRR currently applies to systems of all sizes and is intended to help utilities minimize potential health risks associated with recycle, particularly associated with respect to *Giardia* and *Cryptosporidium*. Other contaminants of concern in the recycle stream include suspended solids (turbidity), dissolved metals (especially iron and manganese), and dissolved organic carbon. Plants that control recycle will also help minimize operational problems.

Prior to the FBRR, no USEPA regulation governed recycle. Regulations within the United States regarding recycle had been established by the states, if at all. State regulatory approaches varied from a requirement of equalization of two backwashes in Illinois to 80 percent solids removal prior to recycle and maintaining recycle flows at less than 10 percent of raw water flow in California.

Key components of the FBRR include (1) recycle must reenter the treatment process *prior* to primary coagulant addition, (2) direct filtration plants must report their recycle practices to the state and may need to treat their recycle streams, and (3) a self-assessment must be done at those plants that use direct recycle (i.e., no separate equalization and/or treatment of recycle stream) and that operate fewer than 20 filters. The goal of the self-assessment is to determine if the design capacity of the plant is exceeded due to recycle practices.

Ground Water Rule (GWR)

USEPA promulgated the final GWR in October 2006 to reduce the risk of exposure to fecal contamination that may be present in PWSs that use groundwater sources. The rule establishes a risk-targeted strategy to identify groundwater systems that are at high risk for fecal contamination. The GWR also specifies when corrective action (which may include disinfection) is required to protect consumers who receive water from groundwater systems from bacteria and viruses.

A sanitary survey is required, by the state primacy agency, at regular intervals depending on the condition of the water system as determined in the initial survey. Systems found to be at high risk for fecal contamination are required to provide 4-log inactivation of viruses. Increased monitoring for fecal contamination indicators may be required by the regulatory authority.

Total Coliform Rule (TCR) and Revised Total Coliform Rule (RTCR)

The objective of the TCR is to promote routine surveillance of distribution system water quality to search for fecal matter and/or disease-causing bacteria. All points in a distribution system cannot be monitored, and complete absence of fecal matter and disease-causing bacteria cannot be guaranteed. The TCR is an attempt to persuade water utilities to implement monitoring programs sufficient to verify that public health is being protected as much as possible, as well as allowing utilities to identify any potential contamination problems in their distribution system. The rule requires monthly sampling at each distribution sampling point.

The TCR, and the RTCR that was finalized in 2013, impact all PWSs. The RTCR requires PWSs that are vulnerable to microbial contamination to identify and fix problems. The RTCR also established criteria for systems to qualify for and stay on reduced monitoring, thereby providing incentives for improved water system operation.

The RTCR rule established an MCLG and an MCL for *Escherichia coli* (*E. coli*) and eliminated the MCLG and MCL for total coliform, replacing it with a treatment technique for coliform that requires assessment and corrective action. The rule establishes an MCLG and an MCL of zero for *E. coli*, a more specific indicator of fecal contamination and potentially harmful pathogens than total coliform. USEPA removed the MCLG and MCL of zero for total coliform. Many of the organisms detected by total coliform methods are not of fecal origin and do not have any direct public health implication.

Under the treatment technique for coliform, total coliform serves as an indicator of a potential pathway of contamination into the distribution system. A PWS that exceeds a specified frequency of total coliform occurrence must conduct an assessment to determine if any sanitary defects exist and, if found, correct them. In addition, a PWS that incurs an *E. coli* MCL violation must conduct an assessment and correct any sanitary defects found.

The RTCR also changed monitoring frequencies. It links monitoring frequency to water quality and system performance and provides criteria that well-operated small systems must meet to qualify and stay on reduced monitoring. It also requires increased monitoring for high-risk small systems with unacceptable compliance history and establishes some new monitoring requirements for seasonal systems such as state and national parks.

The revised rule eliminated monthly public notification requirements based only on the presence of total coliforms. Total coliforms in the distribution system may indicate a potential pathway for contamination but in and of themselves do not indicate a health threat. Instead, the rule requires public notification when an *E. coli* MCL violation occurs, indicating a potential health threat, or when a PWS fails to conduct the required assessment and corrective action.

The rule requires that PWSs collect total coliform samples at sites representative of water quality throughout the distribution system according to a written plan approved by the state or primacy agency. Samples are collected at regular intervals monthly. Positive total coliform samples must be tested for *E. coli*. If any positive total coliform sample is also positive for *E. coli* the state must be notified by the end of the day on which the result was received. Repeat samples are required within 24 hours of any total coliform–positive routine sample. Three repeat samples are required, one at the site of the positive sample and one within five service taps both upstream and downstream of the positive site. Any positive total coliform samples must be tested for *E. coli*. Any positive *E. coli* (EC+) samples must be reported by the end of the day. Any positive total coliform (TC+) samples require another set of repeat samples.

A Level 1 or Level 2 sanitary assessment and corrective action is triggered to occur within 30 days if there is indication of coliform contamination. A Level 1 assessment by the PWS is triggered if more than 5 percent of the routine/repeat monthly samples (if at least 40 are required) are total coliform positive or a repeat sample is not taken for a total coliform positive result. A Level 2 assessment conducted by the state or its representative is triggered if the PWS has an *E. coli* violation or repeated Level 1 assessment triggers.

Major violations of the RTCR are MCL violations and treatment technique violations. A PWS will receive an *E. coli* MCL violation when there is any combination of an EC+ sample result with a routine/repeat TC+ or EC+ sample result, as follows:

E. coli MCL Violation Occurs With the Following Sample Result Combination

Routine	Repeat
EC+	TC+
EC+	Any missing sample
EC+	EC+
TC+	EC+
TC+	TC+ (but no *E. coli* analysis)

A PWS will receive a treatment technique violation given any of the following conditions:

- Failure to conduct a Level 1 or Level 2 assessment within 30 days of a trigger.
- Failure to correct all sanitary defects from a Level 1 or Level 2 assessment within 30 days of a trigger or in accordance with the state-approved time frame.
- Failure of a seasonal system to complete state-approved start-up procedures prior to serving water to the public.

Lead and Copper Rule (LCR)

The objective of the LCR is to control corrosiveness of the finished water in drinking water distribution systems to limit the amount of lead (Pb) and copper (Cu) that may be leached from certain metal pipes and fittings in the distribution system. Of particular concern are pipes and fittings connecting the household tap to the distribution system service line at individual homes or businesses, especially because water can remain stagnant in these service lines for long periods of time, increasing the potential to leach Pb, Cu, and other metals. Although the utility is not responsible for maintaining and/or replacing these household connections, they are responsible for controlling pH and corrosiveness of the water delivered to the consumers.

Details of the LCR include the following:

- The LCR became effective December 7, 1992.
- The action level for Pb is 0.015 mg/L and for Cu is 1.3 mg/L.
- A utility is in compliance at each sampling event (frequency discussed below) when <10 percent of the distribution system samples are above the action levels for Pb and Cu (i.e., 90th percentile value for sampling event must be below action level).
- Utilities found not to be in compliance must modify water treatment until they are in compliance. The term *action level* is used rather than *MCL* because noncompliance (i.e., exceeding an action level) triggers a need for modifications in treatment.
- Additional revisions to the LCR are under evaluation; check with your local regulatory agency for additional clarifications.

After identifying sampling locations and determining initial tap water Pb and Cu levels at each of these locations, utilities must also monitor other water quality parameters (WQPs) at these same locations as needed to monitor and evaluate corrosion control characteristics of treated water. The only exemptions from analysis of these WQPs are systems serving less than 50,000 people for which Pb and Cu levels in initial samples are below action levels.

Pb, Cu, and WQPs are initially collected at 6-month intervals; this frequency can be reduced if action levels are not exceeded and optimal water treatment is maintained. Systems that are in noncompliance and are performing additional corrosion-control activities must continue to monitor at 6-month intervals, plus they must collect WQPs from distribution system entry points every 2 weeks.

Each utility must complete a survey and evaluate materials that comprise their distribution system, in addition to using other available information, to target homes that are at high risk for Pb/Cu contamination.

Revisions to the LCR were enacted in 2007. These clarifications to the existing rule were made in seven areas:

- Minimum number of samples required
- Definitions for compliance and monitoring periods
- Reduced monitoring criteria
- Consumer notice of lead tap water monitoring results
- Advanced notification and approval of long-term treatment changes
- Public education requirements
- Reevaluation of lead service lines

Consult your local regulatory agency for those revisions that are applicable to your system.

Chemical Contaminant Rules

The Phase I, II, and V regulations, known as the Chemical Contaminant Rules, were finalized in 1989, 1992, and 1995, respectively, and include various inorganic and organic contaminants. Sampling and reporting frequency vary with constituent, though sampling is typically required once every 3 years after the initial sampling period. Variances or waivers are possible for a number of constituents based on analytical results and/or a vulnerability assessment.

Public Notification Rule

USEPA has implemented a regulation called the *Public Notification Rule*. This rule is separate from the Consumer Confidence Report (CCR) Rule. The Public Notification Rule includes requirements for reporting certain water quality monitoring violations and other water quality incidents, as well as requirements for the timing, distribution, and language of the public notices. For example, the Public Notification Rule includes requirements that some incidents be reported within 24 hours, others within 30 days, and others included as part of the annual CCR. Some of these reporting requirements are more stringent than those currently required by USEPA. The regulation also includes requirements regarding how notices are to be distributed/broadcast (i.e., TV, radio, newspaper, hand delivery, regular mail, etc.), the format of the notices, the wording of certain items in the notice, and the need to include information in languages other than English.

Public notification according to the rule might include the following:

- Templates, or model notices, to be available for adaptation for certain potential incidents.
- Consolidated and updated lists of phone numbers and contacts for government (local, county, state), regulatory agencies, hospitals, radio and TV, newspapers, etc., that should be contacted per requirements of the Public Notification Rule
- Checklists and flow diagrams outlining activities that would need to be completed for certain potential events outlined in the regulation
- Identification of key personnel and what their roles and responsibilities would be to respond as required by the regulation
- A plan to periodically review and update all lists, templates, and other aspects of a response plan every year or when/if the Public Notification Rule is modified by future federal or state regulations

Unregulated Contaminant Monitoring Rule (UCMR)

The 1996 amendments to the SDWA require USEPA to establish criteria for a monitoring program for currently unregulated contaminants to generate data that USEPA can use to evaluate and prioritize contaminants that could potentially be regulated in the future. USEPA has developed three cycles of the UCMR:

1. UCMR1 in 1999
2. UCMR2 in 2007
3. UCMR3 in 2012

Failing to (1) perform required sampling and analysis, (2) use the appropriate analytical procedures, or (3) report these results is a violation of the UCMR. However, the numerical results of these analytical efforts cannot result in a violation because none of the constituents in the UCMR are currently regulated (i.e., no MCLs, action levels, or other standards apply).

Although the UCMR contaminants have no standards associated with them, the data from this monitoring will need to be reported in the annual CCR. Therefore, the CCR will need to address implications of any constituents found above detection limits. Reporting UCMR results in the CCR would also fulfill the notification requirements for "unregulated contaminants" included in the recently promulgated Public Notification Rule.

Note that the UCMR is an ongoing part of the regulatory development process that will be repeated every 5 years. Utilities will be performing similar mandatory sampling for a new list of constituents every 5 years.

UCMR3 was signed by USEPA Administrator Lisa P. Jackson on April 16, 2012. As finalized, UCMR3 requires monitoring for 30 contaminants using USEPA and/or consensus organization analytical methods during 2013–2015. Together, USEPA, states, laboratories, and PWSs will participate in UCMR3.

Operator Certification

Amendments to the 1996 SDWA required USEPA to develop national guidance for operator certification. The final rule was published on February 5, 1999, and became effective on February 5, 2001. State operator certification programs were required to address nine baseline standards, including operator qualifications, certification renewal, and program review. Indirect impacts of the rule on most water utilities include availability of Drinking Water State Revolving Fund (DWSRF) money and perhaps some slight modifications in paperwork/record-keeping requirements.

Arsenic MCL

The MCL for arsenic was reduced from 50 μg/L to 10 μg/L in the *Federal Register* published on January 22, 2001. This was the second time USEPA has established an MCL that was higher than the technically feasible level (3 μg/L), with the first being the uranium rule in 2000. The original SDWA required the MCL to be set as close to the health goal (zero for arsenic and all other suspected carcinogens) as technically feasible. Amendments to the SDWA allowed USEPA the discretion to set the MCL above the technically feasible level.

The final rule, including the revised MCL, became effective 3 years after the rule was published.

Radionuclides Rule

The Radionuclide Rule was published in December 2000. In the final rule, USEPA maintained the gross alpha MCL at 15 pCi/L MCL, 4 mrem/yr for beta emitters, 4 mrem/yr for photon emitters, and 5 pCi/L for combined radium 226 and 228 isotopes, and an MCL for uranium of 30 μg/L.

Analytical Methods

Each of the individual USEPA regulations contains its own information regarding analytical methods approved for compliance monitoring. These and other approved analytical methods are compiled in a final rule titled "Analytical Methods for Chemical and Microbiological Contaminants and Revisions to Laboratory Certification Requirements," published December 1, 1999. These analytical methods were approved for compliance monitoring effective January 3, 2000. The USEPA-approved methods include analytical procedures developed by USEPA, plus procedures developed by others that USEPA endorses, including specific procedures developed by the American Society for Testing and Materials (ASTM) and some specific procedures included in *Standard Methods for the Examination of Water and Wastewater*, published jointly by the American Public Health Association (APHA), AWWA, and the Water Environment Federation (WEF).

Currently, only approved analytical methods can be used for compliance monitoring. In the future, USEPA hopes to implement a performance-based measurement system that will allow utilities to use alternative screening methods instead of requiring only USEPA-approved reference methods. The 1996 SDWA amendments require USEPA to review new analytical methods that may be used for the screening and analysis of regulated contaminants. After this review, USEPA may approve methods that may be more accurate or cost-effective than established methods for compliance monitoring. These screening methods are expected to provide flexibility in compliance monitoring and may be better and/or faster than existing analytical methods.

The approval of new drinking water analytical methods can be announced through an expedited process in the *Federal Register*. This allows laboratories and water systems more timely access to new alternative testing methods than the traditional rule-making process. If alternate test procedures perform the same as or better than the approved methods, they can be considered for approval using the expedited process.

Study Questions

1. If a water system collects at least 40 samples per month for the analyses of total coliforms, what percentage of total coliform positive samples are acceptable for the system to remain in compliance with the maximum contaminant level for total coliforms?
   a. No more than 2%
   b. No more than 3%
   c. No more than 4%
   d. No more than 5%

2. Water systems are required to achieve at least _________ removal and/or inactivation of viruses between a point where the raw water is not subject to recontamination by surface water runoff and a point downstream before or at the first customer.
   a. 2 log
   b. 2.5 log
   c. 3 log
   d. 4 log

3. Continuous chlorine residual monitoring is required where the water enters the distribution system under the Surface Water Treatment Rule when the
   a. population served is >3,300 people.
   b. population served is >10,000 people.
   c. number of taps is >1,000.
   d. number of taps is >2,500.

4. A community water system must post the Consumer Confidence Report on a publicly accessible website if it serves more than
   a. 10,000 people.
   b. 25,000 people.
   c. 50,000 people.
   d. 100,000 people.

5. Public water systems that cannot meet the required removal of total organic compounds (TOCs) can comply if the source-water TOC level is ________, calculated quarterly as a running annual average.
   a. <1.0 mg/L
   b. <2.0 mg/L
   c. <2.5 mg/L
   d. <5.0 mg/L

6. The Surface Water Treatment Rule does not specify which of the following treatment technology?
   a. Conventional treatment
   b. Membranes
   c. Slow sand filtration
   d. Chloramine disinfection

7. A regulated organic contaminant concentration that exceeds the MCL
   a. is a Tier 1 violation requiring public notification within 24 hours.
   b. is a Tier 2 violation requiring public notification within 30 days.
   c. is a Tier 3 violation requiring public notification within 12 months.
   d. does not require public notification.

8. Operators must balance satisfying regulations on the amount of certain chlorinated organic compounds with
   a. controlling water hardness.
   b. meeting disinfection limits.
   c. keeping turbidity below 0.1 ntu.
   d. reducing iron levels below the MCL.

9. A water system using conventional filtration treatment does not have to use enhanced coagulation to achieve the total organic carbon (TOC) percent removal if
   a. the total trihalomethanes running annual average is less than or equal to 0.030 mg/L.
   b. the sum of haloacetic acids is less than or equal to 0.040 mg/L.
   c. the treated water quarterly running average TOC is less than 4.0 mg/L.
   d. the treated water quarterly running average TOC is less than 2.0 mg/L.

10. Regulations that govern US water supply and treatment are developed by the US Environmental Protection Agency under what act?

11. According to the US Environmental Protection Agency's primary drinking water standards, what is the maximum contaminant level goal for copper?

12. According to the US Environmental Protection Agency's primary drinking water standards, what is the maximum contaminant level for alpha particles?

13. What term refers to the highest level of a disinfectant allowed in drinking water?

14. What is the maximum contaminant level for arsenic?

Chapter 2
Operator Math

Piezometric Surface and Hydraulic Grade Line

Many important hydraulic measurements are based on the difference in height between the free water surface and some point in the water system. The free water surface is the surface of water that is in contact with the atmosphere. The piezometric surface can be used to locate the free water surface in a container, where it cannot be observed directly.

Piezometric Surface

If you connect an open-ended tube (similar to a straw) to the side of a tank or pipeline, the water will rise in the tube to indicate the level of the water in the tank. Such a tube, shown in Figure 2-1, is called a piezometer, and the level of the top of the water in the tube is called the piezometric surface. If the water-containing vessel is not under pressure (as is the case in Figure 2-1), the piezometric surface will be the same as the free water surface in the vessel, just as it would be if a soda straw (the piezometer) were left standing in a glass of water.

If the tank and pipeline are under pressure, as they often are, the pressure will cause the piezometric surface to rise above the level of the water in the tank. The greater the pressure, the higher the piezometric surface (Figure 2-2).

Notice in Figure 2-2A that the free water surface shown by the piezometer is the same level as the water surface in the tank; but once a pressure is applied, as in Figure 2-2B and 2-2C, the free water surface rises above the level of the tank water surface.

The applied pressure caused by a piston in Figure 2-2B and 2-2C can also be caused by water standing in a connected tank at a higher elevation. For example, Figure 2-3 shows that the pressure caused by water in an elevated storage tank

free water surface
The surface of water that is in contact with the atmosphere.

piezometric surface
An imaginary surface that coincides with the level of the water in an aquifer, or the level to which water in a system would rise in a piezometer.

piezometer
An instrument for measuring pressure head in a conduit, tank, or soil, by determining the location of the free water surface.

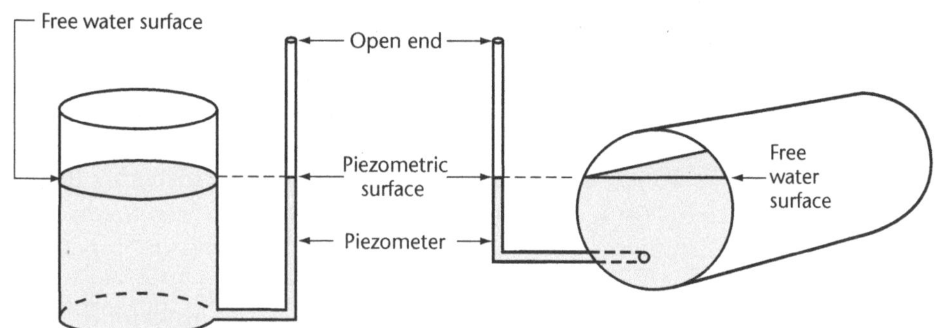

Figure 2-1 Piezometer and piezometric surface

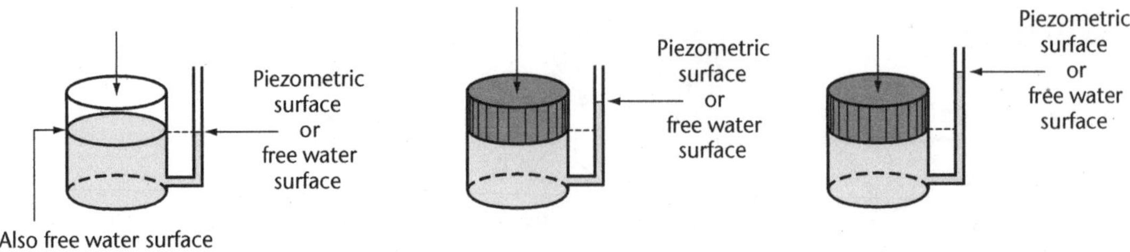

Figure 2-2 Piezometric surface varying with pressure

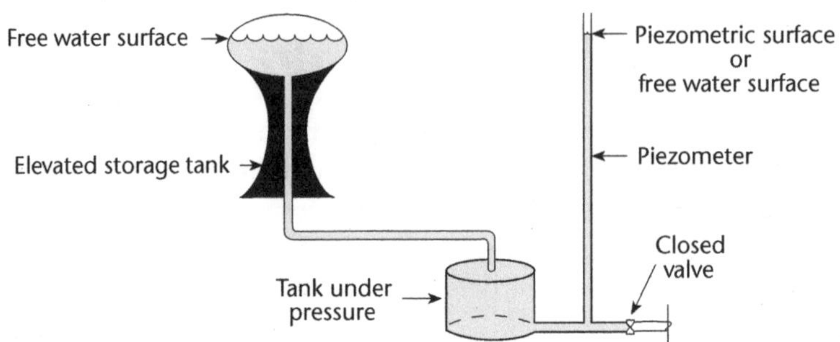

Figure 2-3 Piezometric surface caused by elevated tank

is transmitted down the standpipe, through the pipeline, into a closed, low-level tank, and into the piezometer. The pressure causes the water to rise in the piezometer to the height of the water surface in the storage tank. Notice that the relationship between the tank under pressure and the piezometric surface shown in Figure 2-3 is very similar to the relationship between the pressurized tank and the piezometric surface in Figure 2-2C. In both cases, the piezometric surface is higher than the surface of the water in the pressurized tank to which the piezometer is connected. Note, however, that Figure 2-3 is also similar to Figure 2-2A, in that the piezometric surface in both figures is ultimately at the same level as the free water surface in the open-top tank.

To illustrate the principles of piezometric surfaces (free water surfaces), locate the piezometric surface (free water surface) in Figures 2-4A and 2-5A. The answers are shown in Figures 2-4B and 2-5B. In each case, the piezometric surface is the same as the water surface in the main body of water. This is true no matter where the piezometer is connected and no matter what slope or shape the piezometer takes.

So far only the piezometric surface for a body of standing water (static water) has been considered. The piezometers have shown that the water always rises to the water level of the main body of water, *but only when the water is standing still.*

Changes in the piezometric surface occur when water is flowing. Figure 2-6 shows an elevated storage tank feeding a distribution system pipeline. When the valve is closed (Figure 2-6A), all the piezometric surfaces are the same height as the free water surface in storage. When the valve opens and water begins to flow (Figure 2-6B), the piezometric surfaces *drop*. The farther along the pipeline, the lower the piezometric surface, because some of the pressure is used up keeping the water moving over the rough interior surface of the pipe. The pressure that is lost (called head loss) is no longer available to push water up in the piezometer. As water continues down the pipeline, less and less pressure is left.

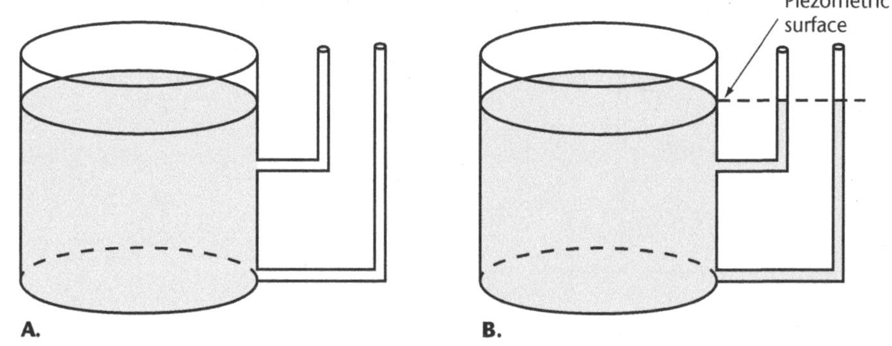

Figure 2-4 Schematic 1 of piezometric surface

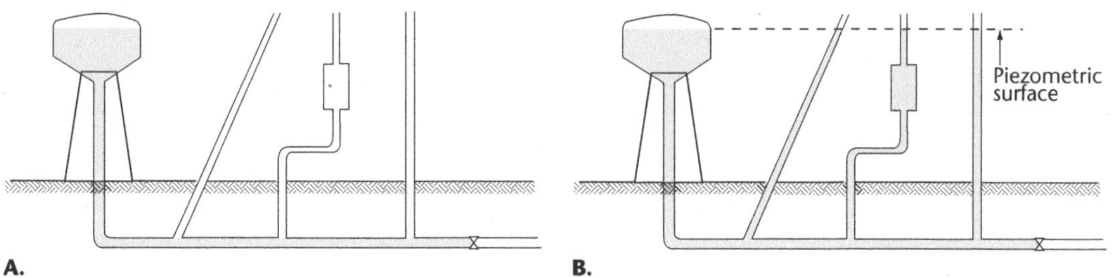

Figure 2-5 Schematic 2 of piezometric surface

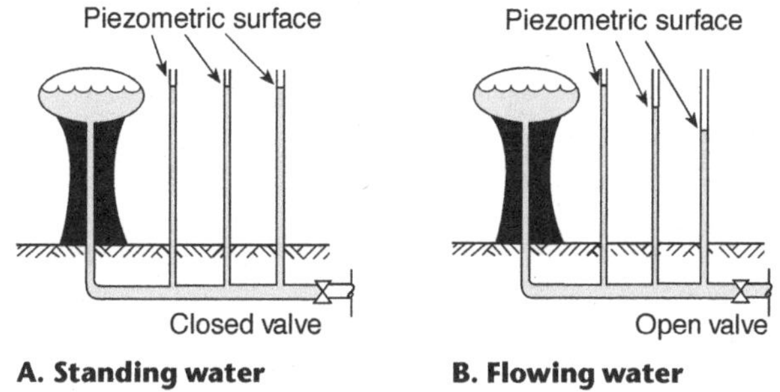

Figure 2-6 Piezometric surface changes when water is flowing

Hydraulic Grade Line

The hydraulic grade line (HGL) is the line that connects all the piezometric surfaces along the pipeline. It is important from an operating standpoint because it can be used to determine the pressure at any point in a water system.

To better understand HGL, you should know how the HGL is located and drawn. In this section, two techniques are discussed:

- Locating HGLs from piezometric surface information
- Locating HGLs from pressure gauge information

hydraulic grade line (HGL)
A line (hydraulic profile) indicating the piezometric level of water at all points along a conduit, open channel, or stream. In an open channel, the HGL is the free water surface.

static water system
The description of a water system when water is not moving through the system.

Locating HGLs From Piezometric Surface Information

First look at how to find the HGL of a static water system (a system in which the water is not moving). The pipeline in Figure 2-7 is fitted with four piezometers. With the valve closed, water rises in three of them, up to the free water surface. To find the HGL, draw a horizontal line from the free water surface through

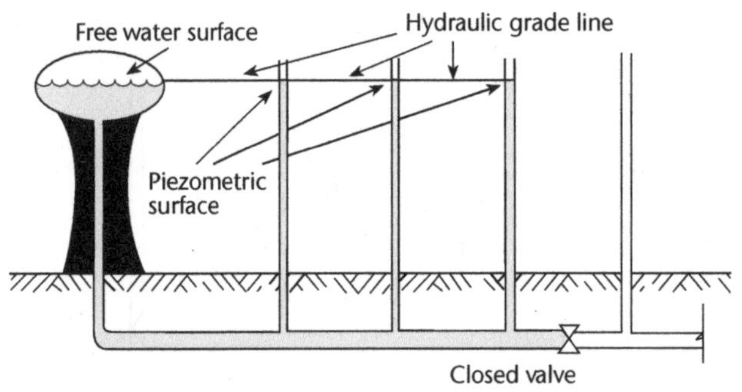

Figure 2-7 HGL of static water system

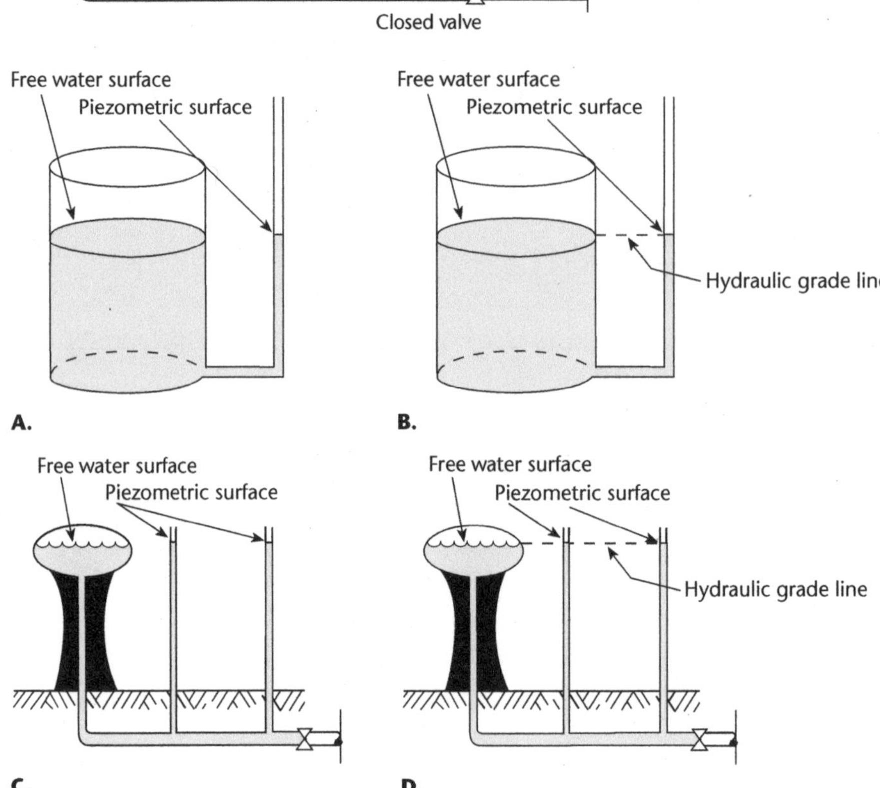

Figure 2-8 Schematics for static water conditions

the piezometric surfaces, as shown. This demonstrates two important facts about HGLs for *static* water systems:

- The static HGL is always horizontal.
- The static HGL is always at the same height as the free water surface.

To illustrate these principles, locate and draw the HGL for the two static water conditions shown in Figures 2-8A and 2-8C. First, the HGL will pass through the piezometric surfaces. Also, because the water is static, the following principles apply:

- The HGL will be horizontal.
- The HGL will be at the free water surface.

Using this information, you can draw the HGL by connecting the piezometric surfaces to the free water surface of the reservoir with a horizontal line. The resulting HGLs are shown in Figures 2-8B and 2-8D.

dynamic water system
The description of a water system when water is moving through the system.

Now consider how to find the HGL of a dynamic water system (a system in which the water is in motion). Figure 2-9 shows water moving from an elevated storage tank into the distribution system. As before, draw the HGL by connecting the free water surface and all the piezometric surfaces. The result can be one straight line, as shown in Figure 2-9, or it can be a series of connected straight lines at different angles, as shown in Figure 2-10.

The dashed line in Figure 2-9 shows the static HGL. The vertical distance between the static HGL and the dynamic HGL at any point along the pipeline is a measure of the amount of pressure the water has used flowing to that point.

Figure 2-10 illustrates how the HGL is established for a typical water transmission line. Notice in this case that the HGL is a series of straight lines at different slopes. There is more head loss wherever the flow enters a smaller-diameter pipe, as indicated by a downward change in HGL slope. Slope changes also show pressure loss that occurs as a result of factors, including flow rate, pipe material, and pipe age and roughness. The changing HGL indicates the combined effect of these factors. As in Figure 2-9, the dashed line indicates where the HGL would be if the water were static. This would be the condition if, for example, flow was stopped by a closed valve just outside the treatment plant.

The concept of hydraulic grade line for dynamic systems also applies to artesian wells. When a well is drilled into an artesian aquifer, as shown by wells *A* and *C* in Figure 2-11, the water will rise in the well, with the well acting as a piezometer. The recharge area of the artesian aquifer acts like the elevated storage tank shown in Figure 2-9, and the rest of the aquifer and the artesian wells can be compared to the pipeline with the piezometers. The HGL slopes downward as the water moves through the aquifer.

Notice that one of the artesian wells (well C) is flowing. This happens because the wellhead (the opening of the well at the ground surface) is *below* the HGL and there is no pipe to contain the water rising to the piezometric surface. The same thing would happen in the system illustrated in Figure 2-10 if any one of the piezometers did not reach up to the HGL.

One situation you are likely to encounter in water system operations is shown in Figure 2-12: pumping from a lower reservoir to a higher reservoir. With the pump off (as in Figure 2-12A), there are two separate HGLs: the lower one representing the water level in reservoir 1 and the higher one representing the water level in reservoir 2.

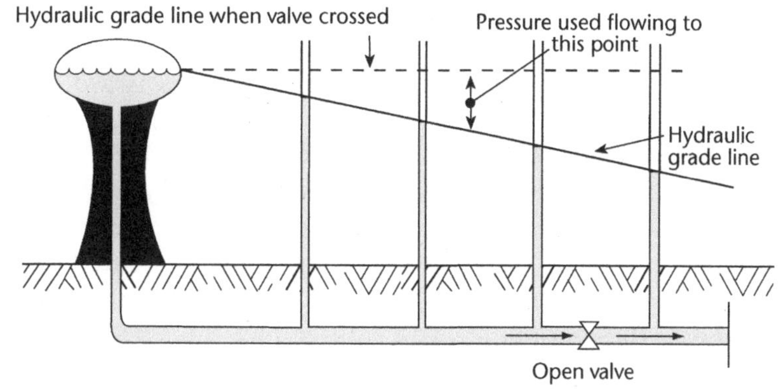

Figure 2-9 HGL of water in motion

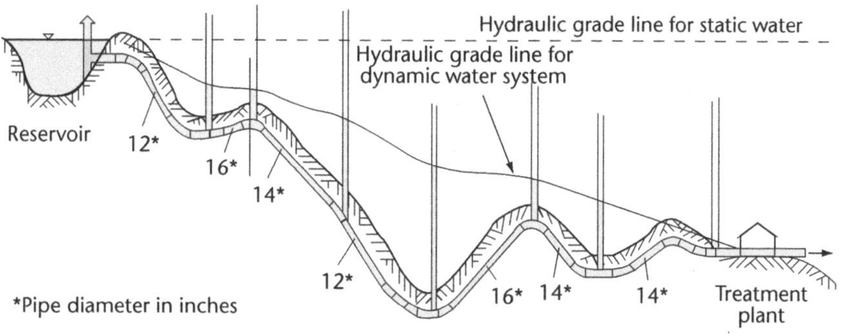

Figure 2-10 HGL of water transmission line

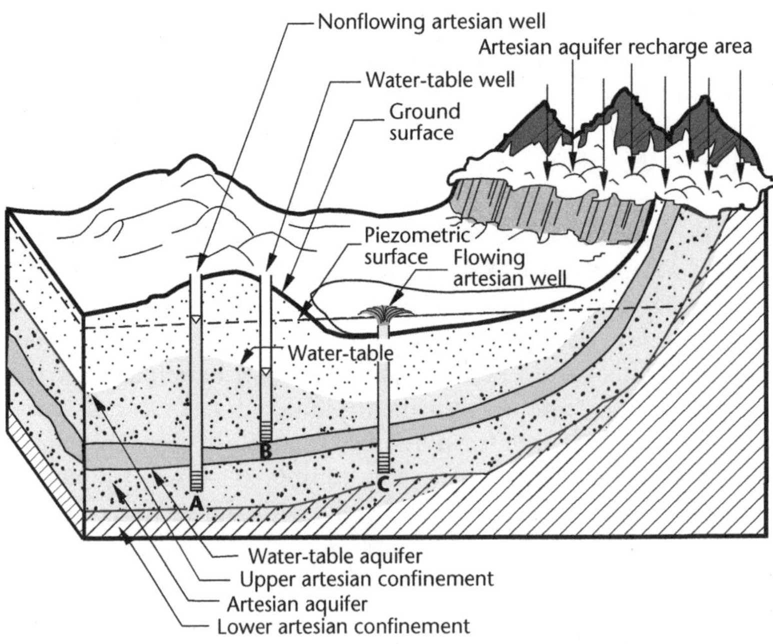

Figure 2-11 HGL and artesian wells

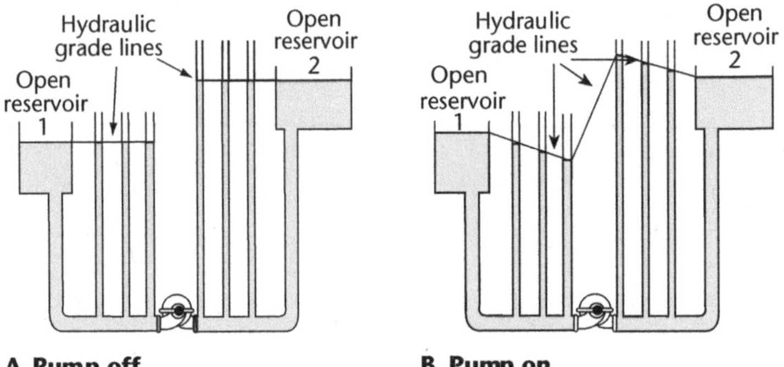

Figure 2-12 Pumping from a lower reservoir to a higher one

When the pump is running (as in Figure 2-12B), notice what happens to the HGL. Starting at reservoir 1, the HGL slopes downward, indicating that the water is losing pressure as it flows from reservoir 1 to the suction, or upstream, side of the pump. The HGL then slopes upward as the water moves through the pump, indicating that pressure has been *added* to the water. Finally, from the discharge side of the pump, the HGL slopes downward again because the water is losing pressure as it flows through the pipe from the pump to reservoir 2. Notice that the HGL ends at the same elevation as the free water surface of reservoir 2.

From these examples, you can see that there are five important basic principles about HGLs for dynamic water systems. Figure 2-13 illustrates these principles for locating an HGL.

- The HGL starts at the same elevation as the free water surface in the upstream reservoir.
- The HGL slopes *downward* in the direction of flow when pressure is used up.
- The HGL slopes upward as water gains pressure by passing through a pump.
- The difference in height between any two points on a downward-sloping HGL shows the pressure used by the water between the two points. Similarly, the difference in height between any two points on an upward-sloping HGL shows the pressure added between those two points.
- The HGL ends at the same elevation as the downstream free water surface.

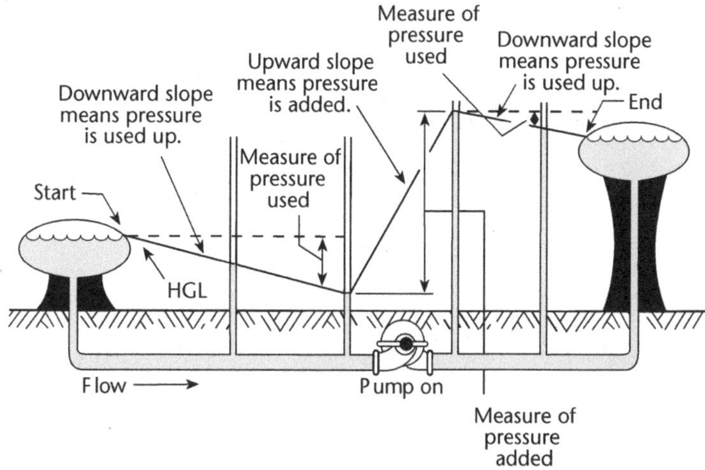

Figure 2-13 Five basic principles of HGLs in dynamic systems

Locating HGLs From Pressure Gauge Information

There are many situations in water transmission, treatment, and distribution for which the use of piezometers for pressure measurement or HGL location is totally impractical. In some cases, for example, piezometers hundreds of feet high would be required. Therefore, in most practical applications, pressure gauge readings are used to locate the HGL. The following example illustrates the procedure:

Example 1

In Figure 2-14A, pressure gauge readings taken along a transmission line are shown in pounds per square inch gauge (psig). From this information, locate and draw the HGL.

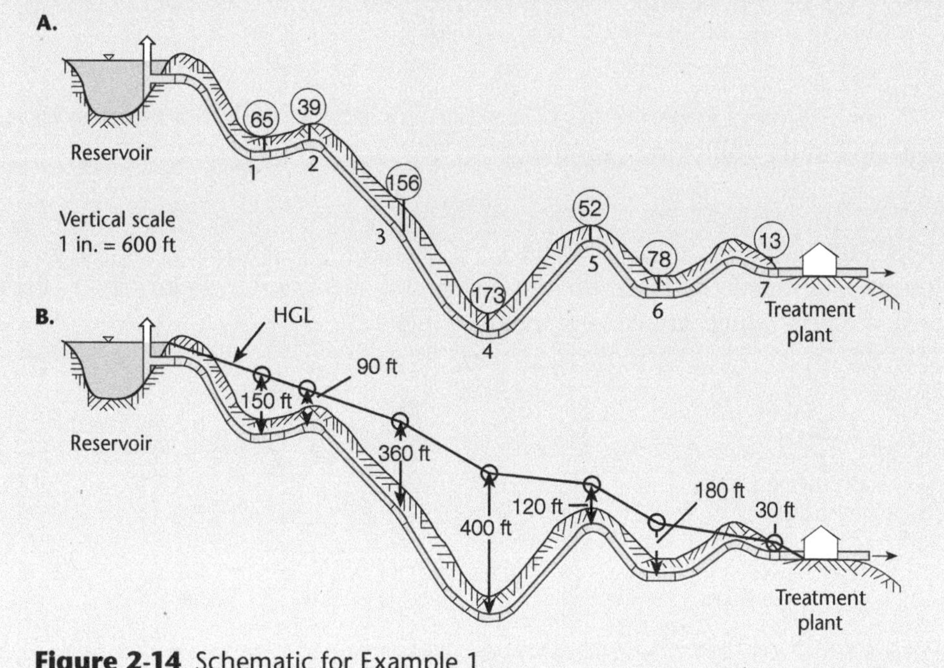

Figure 2-14 Schematic for Example 1

To locate points to scale along the HGL so that a line may be drawn, you must first convert the pressure readings given from psig to pressures in feet. Use the following diagram to make the conversions.

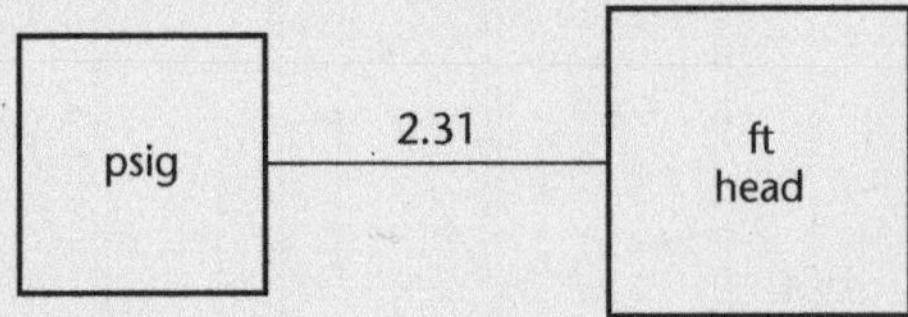

In each conversion from psig pressure to feet of head, you are moving from a smaller box to a larger box. Therefore, you should multiply by 2.31.

Referring to Figure 2-14A, you see that the reading at point 1 is 65 psig. Convert to feet of head as follows:

$$(65 \text{ psig})(2.31 \text{ ft/psig}) = 150.15 \text{ ft}$$

Point 2 has a gauge reading of 39 psig and is converted as follows:

$$(39 \text{ psig})(2.31 \text{ ft/psig}) = 90.09 \text{ ft}$$

Using the same method, you may calculate the pressure in feet for the remaining points:

point 3, 156 psig; pressure = 360.36 ft

point 4, 173 psig; pressure = 399.63 ft

point 5, 52 psig; pressure = 120.12 ft

point 6, 78 psig; pressure = 180.18 ft

point 7, 13 psig; pressure = 30.03 ft

Next, locate the points for pressure in feet *to scale* on the diagram, directly above the pressure gauge locations. You can now draw the HGL by connecting these points, as shown in Figure 2-14B.

Pressure gauges are often the best way to locate the HGL in a pump system, as illustrated in the following example.

Example 2

Locate and draw the HGL for the pump system shown in Figure 2-15. Pressure gauge readings are in kilopascals (gauge).

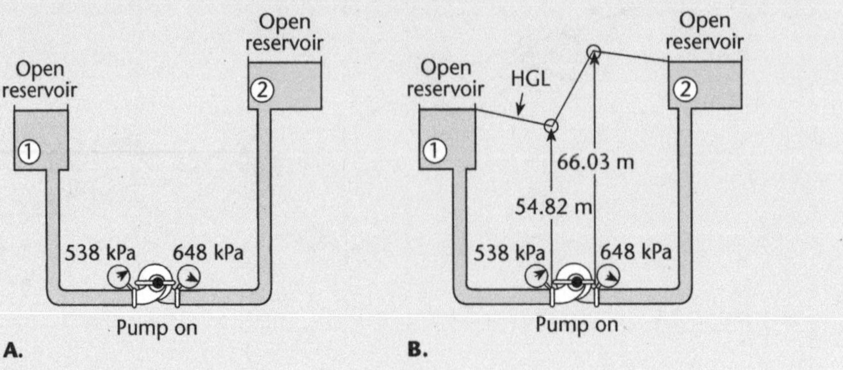

Figure 2-15 Schematic for Example 2

We will convert the gauge pressures to meters of head:

Converting from kilopascals (gauge) to meters of head, you are moving from a smaller box to a larger box. Therefore, multiplication by 0.1019 is indicated.

Referring to Figure 2-15A, you see that the first gauge reading is 538 kPa. Convert to pressure in meters of head as follows:

$$(538 \text{ kPa})(0.1019 \text{ m/kPa}) = 54.82 \text{ m}$$

Then convert the 648-kPa reading to meters of head as follows:

$$(648 \text{ kPa})(0.1019 \text{ m/kPa}) = 66.03 \text{ m}$$

The two points are located (to scale) directly above the gauge points, and the HGL is drawn, as shown in Figure 2-15B.

WATCH THE VIDEO
Piezometric and Hydraulic Grade (www.awwa.org/wsovideoclips)

Head Loss

Head loss is defined as the amount of energy used up by water moving from one point to another. The two most important categories of head loss are friction head loss and minor head loss.

Friction Head Loss

The inside of a pipe may feel quite smooth, but it is rough from a hydraulic standpoint. The friction caused by water moving over the rough surface causes an energy loss called **friction head loss**.

The amount of friction head loss occurring in a pipe varies, depending on (1) the roughness of the pipe and (2) the velocity of flow through the pipe.

Pipe roughness varies widely depending on the type of pipe material and the condition of the pipe. Plastic is one of the smoothest pipe materials, and steel is one of the roughest. A smooth interior surface results in lower friction loss; a rough interior surface causes high loss.

The condition of the pipe depends on several factors:

- Type of lining
- Age
- Degree of corrosion or scaling
- Slime growths
- Tuberculation
- Obstructions (such as mud, silt, sand, rocks, sticks)

Smooth linings reduce friction loss. The other factors listed increase friction loss and cause friction loss to vary from point to point within the water system.

The velocity of the moving water also affects friction loss. Without exception, as the velocity increases, friction loss increases, even if the pipe roughness stays the same.

head loss
The amount of energy used by water in moving from one location to another.

friction head loss
The head lost by water flowing in a stream or conduit as the result of (1) the disturbance set up by the contact between the moving water and its containing conduit and (2) intermolecular friction.

Some of the factors that cause friction loss, such as corrosion, scaling, slime growths, and obstructions, can be controlled by proper operation, maintenance, and treatment of the pipeline. However, the natural roughness of the pipe material or lining, and the natural roughening of that material or lining with age, are factors over which there is little control. Consequently, pipelines must be designed to operate with the additional head needed to overcome friction head loss.

Friction head loss is one of the simplest head losses to calculate. It can be determined from one of many readily available tables. Perhaps the most commonly used are the tables based on the Hazen–Williams formula, in which a smoothness coefficient C represents the smoothness of varying pipe materials of varying ages; the higher the C value, the smoother the pipe. For example, the C value for cast-iron pipe (CIP) may be as high as 140 for the very best new CIP laid perfectly straight. A C value of 130 is the average value for most new CIP. As CIP ages, the C value drops, on the average, to C = 120 at 5 years, C = 110 at 10 years, C = 100 at 17 years, C = 90 at 26 years, C = 80 at 37 years, and C = 75 or less for pipe 40 or more years old. Table 2-1 lists typical smoothness coefficients.

Table 2-1 Smoothness coefficients for various pipe materials

	C Value	
Type of Pipe	**New**	**10 Years Old**
Asbestos–cement pipe	140	120–130
Cast-iron pipe	130	100
Reinforced and plain concrete pipe	140	120–130
Ductile iron pipe	130	100
Plastic pipe	150	120–130
Steel pipe	110	100

The following examples illustrate friction head loss calculations.

Example 3

A 12-in.-diameter water transmission line, 10,500 ft long, carries water from the supply reservoir to the treatment plant at a flow rate of 2.0 mgd. The steel transmission line has been in service for about 10 years. Use Table 2-2 to determine the total friction head loss in the line.

To use Table 2-2, you need to know two numbers: the flow rate (2.0 mgd in this problem) and the C value. According to the problem, the pipe is steel and about 10 years old. Table 2-1 shows a smoothness coefficient for 10-year-old steel pipe of C = 100.

With this information, enter Table 2-2 in the first column at the flow rate 2,000,000 gpd. Then move across the table until you reach the column for C = 100. The number shown is 7.6. This is the friction head loss, in feet, for every 1,000-ft length of pipeline.

To determine the total friction head loss asked for in the problem, you must determine how many 1,000-ft lengths there are in 10,500 ft:

$$\frac{10{,}500 \text{ ft}}{1{,}000 \text{ ft}} = 10.5$$

Table 2-2 Friction loss factors for 12-in. pipe

Discharge		Velocity	Velocity	Loss of Head, ft per 1,000 ft of length						
gpd	ft³/s	ft/s	Head, ft	C = 140	C = 130	C = 120	C = 110	C = 100	C = 90	C = 80
100,000	0.155	0.20	0.00	0.02	0.02	0.02	0.02	0.03	0.04	0.04
200,000	0.309	0.39	0.00	0.06	0.07	0.08	0.09	0.11	0.13	0.16
300,000	0.464	0.59	0.01	0.12	0.14	0.16	0.19	0.22	0.27	0.34
400,000	0.619	0.79	0.01	0.20	0.24	0.27	0.32	0.38	0.47	0.58
500,000	0.774	0.99	0.02	0.31	0.36	0.41	0.48	0.58	0.71	0.88
600,000	0.928	1.18	0.02	0.44	0.50	0.58	0.68	0.81	0.99	1.23
700,000	1.083	1.38	0.03	0.58	0.66	0.77	0.91	1.08	1.32	1.64
800,000	1.238	1.58	0.04	0.74	0.85	0.99	1.15	1.38	1.68	2.09
900,000	1.392	1.77	0.05	0.92	1.06	1.23	1.45	1.72	2.10	2.61
1,000,000	1.547	1.97	0.06	1.12	1.29	1.50	1.76	2.10	2.57	3.18
1,100,000	1.702	2.17	0.07	1.34	1.54	1.79	2.10	2.50	3.04	3.79
1,200,000	1.857	2.36	0.09	1.58	1.81	2.10	2.47	2.94	3.58	4.45
1,300,000	2.011	2.56	0.10	1.83	2.10	2.43	2.85	3.40	4.14	5.2
1,400,000	2.166	2.76	0.12	2.10	2.40	2.79	3.26	3.90	4.76	5.9
1,500,000	2.321	2.96	0.14	2.39	2.73	3.17	3.71	4.43	5.4	6.7
1,600,000	2.476	3.15	0.15	2.69	3.09	3.58	4.20	5.0	6.1	7.6
1,700,000	2.630	3.35	0.17	3.00	3.45	4.00	4.69	5.6	6.8	8.5
1,800,000	2.785	3.55	0.20	3.33	3.82	4.43	5.2	6.2	7.6	9.4
1,900,000	2.940	3.74	0.22	3.70	4.24	4.92	5.8	6.9	8.4	10.4
2,000,000	3.094	3.94	0.24	4.06	4.65	5.4	6.4	7.6	9.2	11.5
2,200,000	3.404	4.33	0.29	4.85	5.6	6.5	7.6	9.0	10.9	13.7
2,400,000	3.713	4.73	0.35	5.7	6.5	7.6	8.9	10.5	12.8	16.0
2,600,000	4.023	5.12	0.41	6.6	7.6	8.8	10.3	12.3	15.0	18.6
2,800,000	4.332	5.52	0.47	7.6	8.7	10.1	11.9	14.1	17.2	21.5
3,000,000	4.642	5.91	0.54	8.6	9.9	11.5	13.5	16.0	19.4	24.3
3,500,000	5.41	6.89	0.74	11.4	13.2	15.3	17.9	21.3	26.0	32.3
4,000,000	6.19	7.88	0.96	14.5	16.6	19.3	22.6	27.0	33.2	41
4,500,000	6.96	8.87	1.22	18.0	20.6	24.0	28.2	33.6	41.2	51
5,000,000	7.74	9.85	1.50	22.0	25.1	29.2	34.3	41.0	50.0	62
5,500,000	8.51	10.84	1.82	26.5	30.3	35.1	41.4	49.4	60	75
6,000,000	9.28	11.82	2.17	31.1	35.7	41.4	48.8	58	70	88
7,000,000	10.83	13.79	2.96	41.2	47.2	55	65	77	94	116
8,000,000	12.38	15.76	3.86	53	61	71	83	99	121	150
9,000,000	13.92	17.73	4.89	66	75	87	103	122	148	185
10,000,000	15.47	19.70	6.03	81	93	107	126	150	183	228

Because the number of 1,000-ft lengths of steel pipe is 10.5, the total friction head loss is

$$\text{total friction head loss} = (\text{loss per 1,000 ft})(\text{no. of 1,000-ft increments})$$
$$= (7.6 \text{ ft})(10.5)$$
$$= 79.8 \text{ ft}$$

Table 2-2 provides other useful information not requested in Example 3 and can be used to solve for several variables relating to friction loss in 12-in. pipe. For example, the third column gives the velocity for the known flow rate (3.94 ft/s for 2.0 mgd), and the fourth column gives the velocity head (0.24 ft). While Table 2-2 is specific to 12-in. pipe, similar tables exist for other sizes of pipe.

Pumping Rates

The mathematical equation used in pumping rate problems can usually be determined from the verbal statement of the problem:

VERBAL: What is the pumping rate in "gallons *per* minute"?

MATH: $$\text{pumping rate} = \frac{\text{gallons}}{\text{minutes}}$$

VERBAL: What is the pumping rate in "gallons *per* hour"?

MATH: $$\text{pumping rate} = \frac{\text{gallons}}{\text{hours}}$$

The number of gallons pumped during a period can be determined either by reading a flowmeter or by measuring the number of gallons pumped into or out of a tank.

Example 4

The totalizer of the meter on the discharge side of your pump reads in hundreds of gallons. If the totalizer shows a reading of 108 at 1:00 p.m. and 312 at 1:30 p.m., what is the pumping rate expressed in gallons per minute?

The problem asks for pumping rate in *gallons per minute* (gpm), so the mathematical setup is

$$\text{pumping rate} = \frac{\text{gallons}}{\text{minutes}}$$

To solve the problem, fill in the blanks (number of gallons and number of minutes) in the equation. The total gallons pumped is determined from the totalizer readings:

$$\begin{array}{r} 31{,}200 \text{ gal} \\ -\ 10{,}800 \text{ gal} \\ \hline 20{,}400 \text{ gal} \end{array}$$

The volume was pumped between 1:00 p.m. and 1:30 p.m., for a total of 30 min. From this information, calculate the gallons-per-minute pumping rate:

$$\text{pumping rate} = \frac{20{,}400 \text{ gal}}{30 \text{ min}}$$

$$= 680 \text{ gpm pumping rate}$$

Instead of using totalizer readings to calculate the average pumping rate for a period of a few minutes or hours (as in Example 4), you can read the *instantaneous* pumping rate or instantaneous flow rate—the flow or pumping rate at one particular moment—directly from many flowmeters. Other flowmeters require that you perform calculations to determine the instantaneous flow rate.

WATCH THE VIDEO
Pump Rates (www.awwa.org/wsovideoclips)

Pump Heads

Pump head measurements are used to determine the amount of energy a pump can or must impart to the water. These heads are measured in feet. Specific terms are used to describe heads measured at various points and under different conditions of the pump system.

Suction and Discharge

In pump systems, the words *suction* and *discharge* identify the inlet and outlet sides of the pump. As shown in Figure 2-16, the suction side of the pump is the *inlet*, or *low-pressure*, side. The discharge side is the *outlet*, or *high-pressure*, side. Any time a pump term includes the words *suction* or *discharge*, you can recognize immediately which side of the pump system is being discussed.

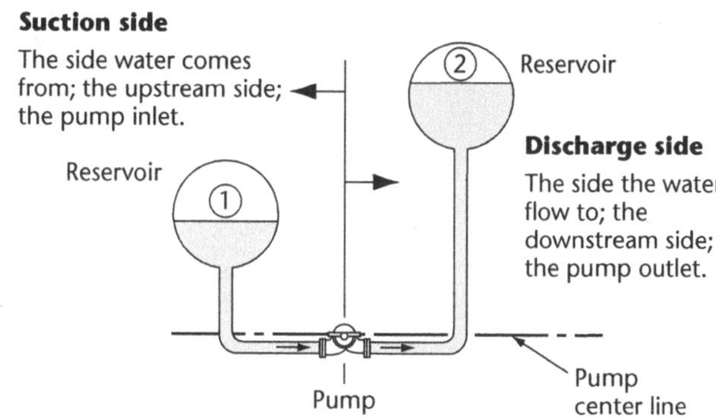

Figure 2-16 Suction and discharge sides of a pump

instantaneous flow rate
A flow rate of water measured at one particular instant, such as by a metering device, involving the cross-sectional area of the channel or pipe and the velocity of the water at that instant.

suction side
The inlet, or low-pressure, side of a pump.

discharge side
The outlet, or high-pressure, side of a pump.

pump center line
An imaginary line through the center of a pump.

static suction head
The difference in elevation between the pump center line and the free water surface of the reservoir feeding the pump. In the measurement of static suction head, the piezometric surface of the water at the suction side of the pump is higher than the pump; otherwise, static suction lift is measured.

static suction lift
The difference in elevation between the pump center line of a pump and the free water surface of the liquid being pumped. In a static suction lift measurement, the piezometric surface of the water at the suction side of the pump is lower than the pump; otherwise, static suction head is measured.

static discharge head
The difference in height between the pump center line and the level of the discharge free water surface.

total static head
The total height that the pump must lift the water when moving it from one point to another. The vertical distance from the suction free water surface to the discharge free water surface.

The **pump center line** is represented by a horizontal line drawn through the center of the pump. This line is important because it is the reference line from which pump head measurements are made.

The terms *suction* and *discharge* help establish the location of the particular pump head measurement. The terms *static* and *dynamic*, however, describe the condition of the system when the measurement is taken. Static heads are measured when the pump is off, and therefore the water is not moving. Dynamic heads are measured with the pump running and water flowing through the system.

Static Heads

Figure 2-17 illustrates the two basic pumping configurations found in water systems. The labeled vertical distances show the four types of static pump head:

- Static suction head
- Static suction lift
- Static discharge head
- Total static head

In a system where the reservoir feeding the pump is higher than the pump (Figure 2-17A), the difference in elevation (height) between the pump center line and the free water surface of the reservoir feeding the pump is termed **static suction head**. But in a system where the reservoir feeding the pump is lower than the pump (as in Figure 2-17B), the difference in elevation between the center line and the free water surface of the reservoir feeding the pump is termed **static suction lift**. Notice that a single system will have either a static head measurement or a static suction lift measurement, but not both.

Static discharge head is defined as the difference in height between the pump center line and the level of the discharge free water surface.

Total static head is the total height that the pump must lift the water when moving it from *reservoir 1* to *reservoir 2*. It is defined precisely as the vertical distance from the suction free water surface to the discharge free water surface. In Figure 2-17A, total static head is found by subtracting static suction head from static discharge head, whereas in Figure 2-17B, it is the sum of the static suction lift and static discharge head.

Static heads can be calculated from measurements of reservoir and pump elevations or from pressure gauge readings taken when the pump is not running.

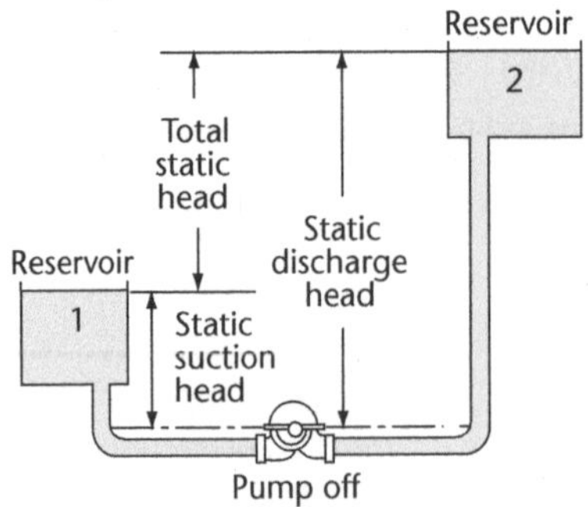

A. Reservoir feeding pump higher than pump

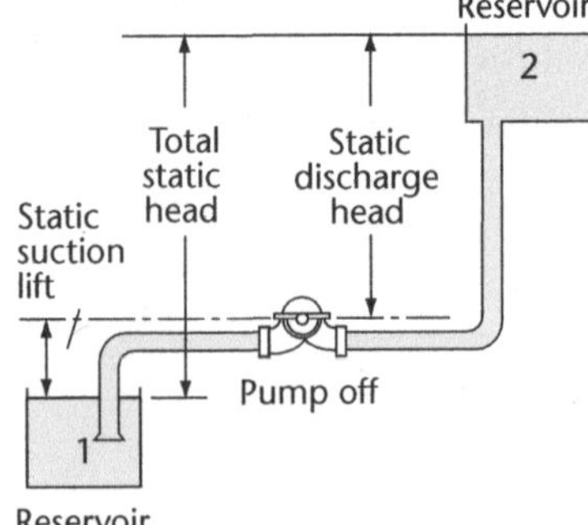

B. Reservoir feeding pump lower than pump

Figure 2-17 Static heads

Dynamic Heads

When water flows through a pipe, the water rubs against the walls of the pipe, and energy is lost to friction. In addition, there is a certain amount of resistance to flow as the water passes through valves, fittings, inlets, and outlets of a piping system. These additional energy losses that the pump must overcome are called *friction head losses* and **minor head losses**.

Let's look at the effect of these additional energy losses on each side of the pump. Figure 2-18 shows a comparison of the head that exists on the suction side of the pump before and after the pump is turned on. Notice that loss of head is caused by friction losses when the pump is turned on. Part of the static suction head (which could otherwise aid the pump in pumping the water up to reservoir 2) is lost because of friction and minor losses as the water moves from reservoir 1 to the pump.

Similarly, friction and minor head losses develop on the discharge side of the pump as the water moves from the discharge side of the pump to reservoir 2. These losses create an additional load or head against which the pump must operate (Figure 2-19). If the pump does not add enough energy to overcome the friction and minor head losses that occur on both the suction and discharge sides of the pump, the pump will not be able to move the water.

Now that you have a basic understanding of the reasons for a difference between static and dynamic conditions, consider the four measurements of dynamic head (shown in Figure 2-20):

- Dynamic suction head
- Dynamic suction lift
- Dynamic discharge head
- Total dynamic head

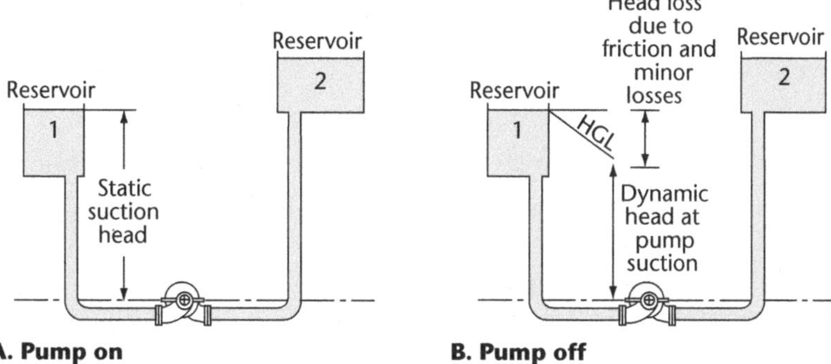

Figure 2-18 Effect of friction and minor head losses on suction head

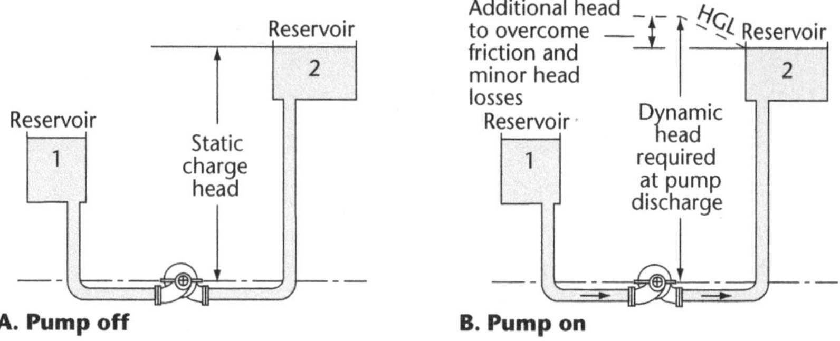

Figure 2-19 Effect of friction and minor head losses on discharge head

minor head loss
The energy losses that result from the resistance to flow as water passes through valves, fittings, inlets, and outlets of a piping system.

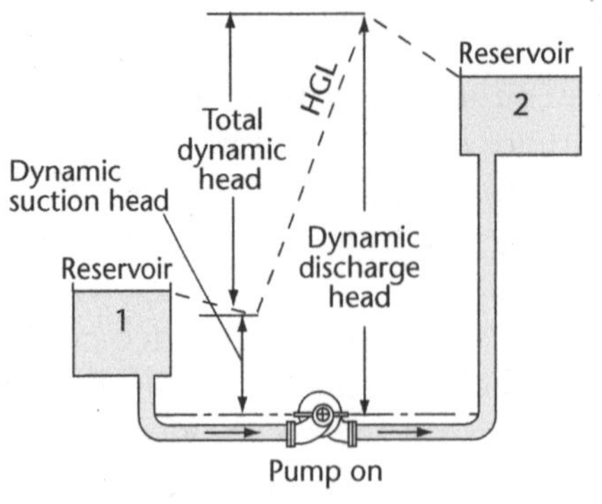

A. Suction side higher than pump

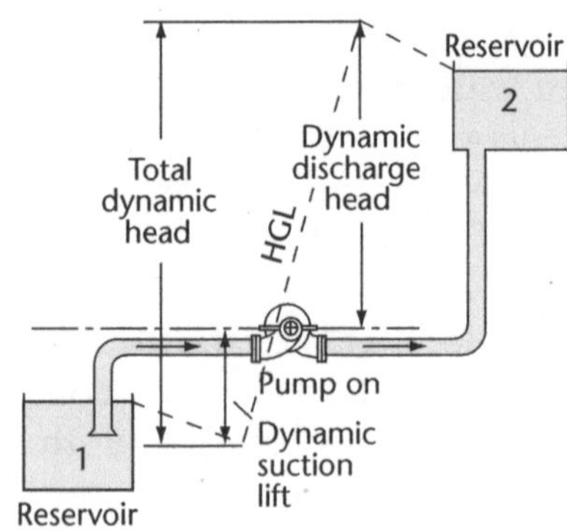

B. Suction side lower than pump

Figure 2-20 Dynamic heads

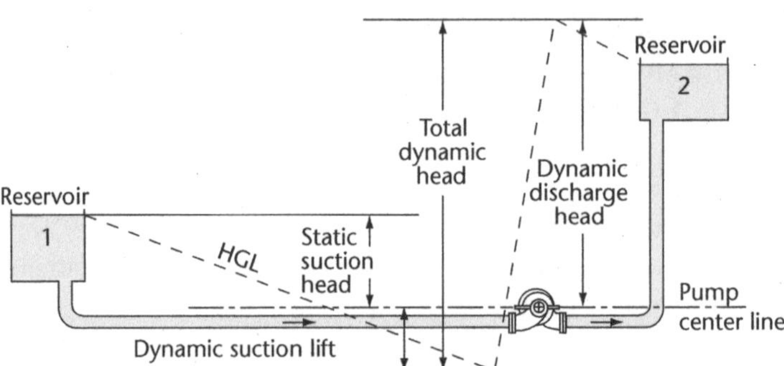

Figure 2-21 Dynamic suction lift for reservoir above pump

dynamic suction head
The distance from the pump center line at the suction of the pump to the point of the HGL directly above it, which exists only when the pump is below the piezometric surface of the water at the pump suction. When the pump is above the piezometric surface, the equivalent measurement is dynamic suction lift.

dynamic suction lift
The distance from the pump center line at the suction of the pump to the point on the HGL directly below it, which exists only when the pump is above the piezometric surface of the water at the pump suction. When the pump is below the piezometric surface, the equivalent measurement is called dynamic suction head.

dynamic discharge head
The difference in height from the pump center line at the pump discharge to the point on the HGL directly above it.

total dynamic head
The difference in height between the HGLs on the discharge side of the pump and on the suction side of the pump. This is a measure of the total energy that a pump must impart to the water to move it from one point to another.

In a system where the HGL on the suction side is higher than the pump (as in Figure 2-20A), the **dynamic suction head** is measured from the pump center line at the *suction* of the pump to the point on the HGL directly *above* it. In a system where the HGL on the suction side is lower than the pump (as in Figure 2-20B), the **dynamic suction lift** is measured from the pump center line at the *suction* of the pump to the point on the HGL directly *below* it.

A single system will have a dynamic suction head or a dynamic suction lift, but not both. The location of the HGL, not of the reservoir, determines which condition exists. As shown in Figure 2-21, it is possible to have a dynamic suction lift condition even though the reservoir feeding the pump is above the pump center line; in such cases, the HGL on the suction side is below the pump center line when the pump is running.

The **dynamic discharge head** is measured from the pump center line at the discharge of the pump to the point on the HGL directly *above* it.

The **total dynamic head** is the difference in height between the HGL on the discharge side of the pump and the HGL on the suction side of the pump. This head is a measure of the total energy that the pump must impart to the water to move it from reservoir 1 to reservoir 2.

Dynamic pump heads can be calculated in either of two ways:

- *By adding friction and minor losses to static head.* Static heads can be determined using existing elevation information; and friction and minor losses can be found by using tables such as Table 2-2.
- *By direct measurement using pressure gauges.* Pressure gauges cannot measure the velocity head component. However, velocity head can often be ignored without significant error.

Horsepower

To understand horsepower, you must first understand the technical meaning of the term work. Work is defined as the operation of a force over a specific distance. For example, suppose a 1-lb object is lifted 1 ft; the amount of work done is measured in foot-pounds (ft-lb):

$$\text{(feet)(pounds)} = \text{foot-pounds}$$

Because it always requires the same amount of work to lift a 1-lb object 1 ft straight up, that amount of work is used as a standard measure: 1 ft-lb. However, work performed on an object of any weight can be measured in foot-pounds (Figure 2-22). Work can be performed in any direction (Figure 2-23). And, as shown in Figure 2-23, work doesn't need to involve lifting; an engine pushing a car along a level highway and a pump pushing water through a level pipeline are both examples of work.

The rate of doing work—that is, the measure of how much work is done in a given time—is called power. To make power calculations, you must know the time required to perform the work. The basic unit for power measurement is foot-pounds per minute (ft-lb/min). Note that this is work performed *per* time. One equation for calculating power in foot-pounds per minute is

$$\left(\begin{array}{c}\text{power in}\\ \text{foot-pounds per minute}\end{array}\right) = \left(\begin{array}{c}\text{head}\\ \text{in feet}\end{array}\right)\left(\begin{array}{c}\text{flow rate in}\\ \text{pounds per minute}\end{array}\right)$$

You will often work with measurements of power expressed in horsepower (hp), which is related to foot-pounds per minute by the conversion equation

$$1 \text{ hp} = 33{,}000 \text{ ft-lb/min}$$

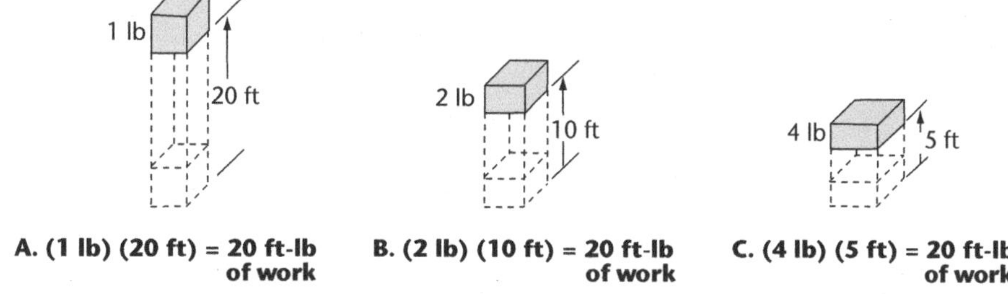

Figure 2-22 Equal amounts of work applied to objects of different weight

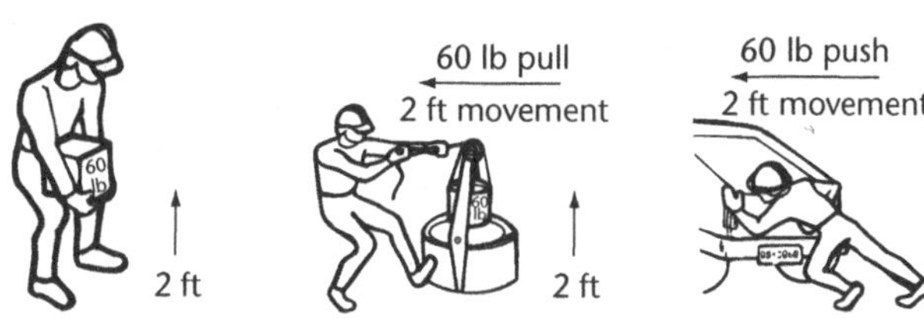

Figure 2-23 Work being performed in different directions

work
The operation of a force over a specific distance.

power (in hydraulics or electricity)
The measure of the amount of work done in a given period of time. The rate of doing work. Measured in watts or horsepower.

When doing horsepower problems using the basic power equation, you must first convert the measurements of head, flow rate, and power into the proper units:

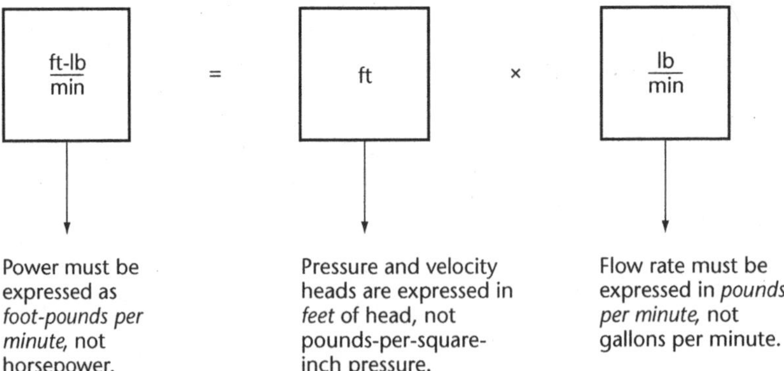

The power calculated in foot-pounds per minute can be converted to horsepower by dividing by 33,000. (The conversion equation is 1 hp = 33,000 ft-lb/min.) The result is called **water horsepower** (whp), because it is the amount of horsepower required to lift the water. Another equation that can be used to calculate water horsepower is

$$\text{whp} = \frac{\text{(flow rate in gallons per minute)(total head in feet)}}{3{,}960}$$

As an alternative, we will also use this equation in our next examples.

Example 5

A pump must pump 2,000 gpm against a total head of 20 ft. What horsepower (water horsepower) will be required to do the work?

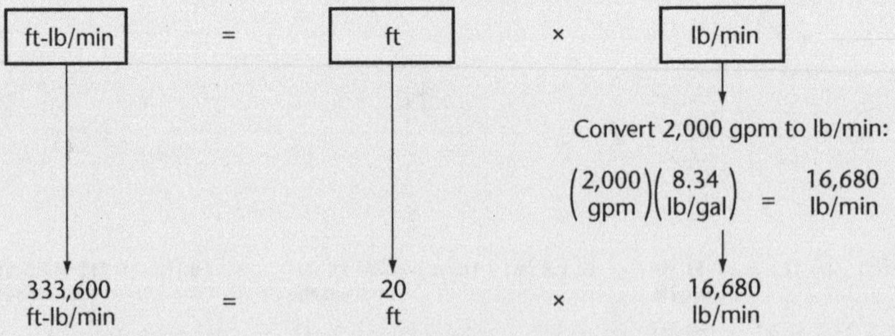

Now convert foot-pounds per minute to horsepower. You know that 1 hp = 33,000 ft-lb/min; in asking, "How much horsepower does this represent?" you are asking, "How many 33,000s are there in 333,600?" Mathematically, this is written as

$$\frac{333{,}600 \text{ ft-lb/min}}{33{,}000 \text{ ft-lb/min/hp}} = 10.11 \text{ hp}$$

As an alternative, you can use the box diagram method to convert foot-pounds per minute to horsepower.

water horsepower (WHP)
The portion of the power delivered to a pump that is actually used to lift water.

You are moving from a larger box to a smaller box. Therefore, division by 33,000 is indicated:

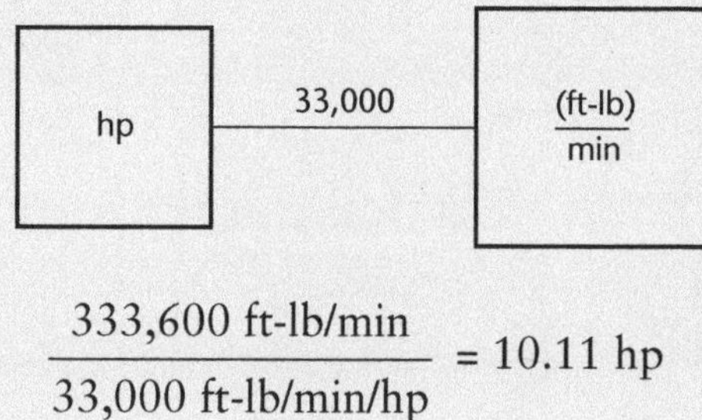

$$\frac{333{,}600 \text{ ft-lb/min}}{33{,}000 \text{ ft-lb/min/hp}} = 10.11 \text{ hp}$$

Example 6

A flow of 450 gpm must be pumped against a head of 50 ft. What is the water horsepower required?

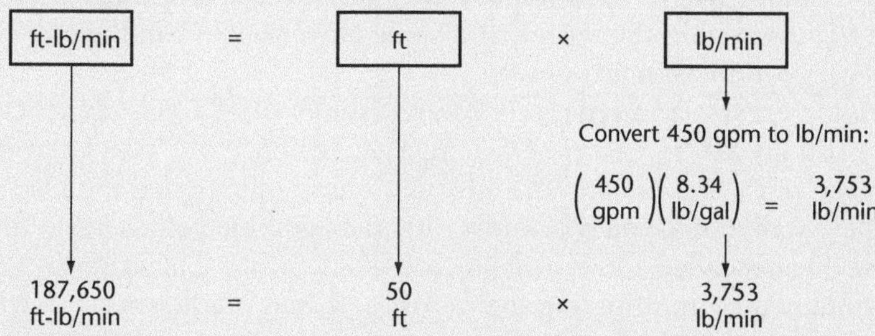

Convert 187,650 ft-lb/min to horsepower (1 hp = 33,000 ft-lb/min) by asking, "How many 33,000s are there in 187,650?" Mathematically, this is written

$$\frac{187{,}650}{33{,}000} = 5.69 \text{ hp}$$

Example 6 Alternative Method

$$\text{hp} = \frac{450 \text{ gpm} \times 50 \text{ ft}}{3{,}960} = 5.68 \text{ hp}$$

WATCH THE VIDEO

Horsepower (www.awwa.org/wsovideoclips)

Efficiency

In the preceding examples, you learned how to calculate water horsepower, the amount of power that must be applied directly to water to move it at a given rate against a given head. Applying the power to the water requires a pump, which in turn is driven by a motor, which is powered by electrical current.

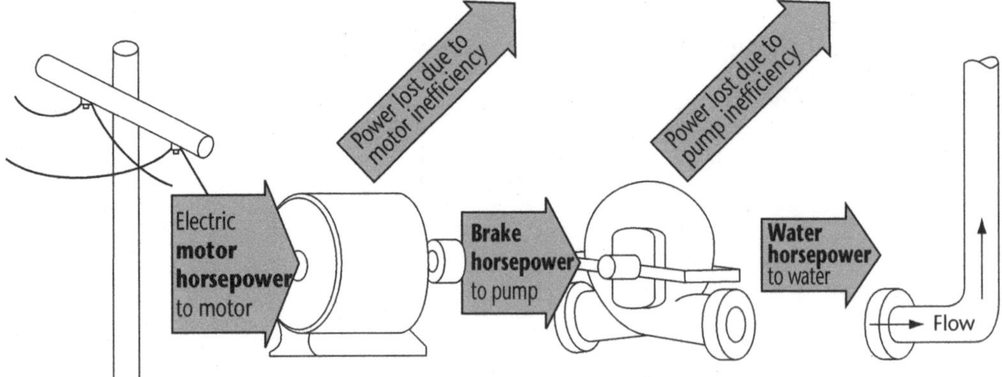

Figure 2-24 Power loss due to motor and pump inefficiency

Neither the pump nor the motor will ever be 100 percent efficient (Figure 2-24). This means that not all of the power supplied by the motor to the pump (called brake horsepower, bhp) will be used to lift the water (*water horsepower*); some of the power is used to overcome friction within the pump. Similarly, not all of the power of the electrical current driving the motor (called motor horsepower, mhp) will be used to drive the pump; some of the current is used to overcome friction within the motor, and some current is lost in the conversion of electrical energy to mechanical power.

Depending on size and type, pumps are usually 50–85 percent efficient, and motors are usually 80–95 percent efficient. The efficiency of a particular motor or pump is given in the manufacturer's information accompanying the unit.

In some installations, you will know only the combined efficiency of the pump and motor. This is called the wire-to-water efficiency. The wire-to-water efficiency is obtained by multiplying the motor and pump efficiencies together. For example, if a motor is 82 percent efficient and a pump is 67 percent efficient, the overall, or wire-to-water, efficiency is 55 percent (0.82 × 0.67 = 0.55).

brake horsepower
The power supplied to a pump by a motor.

motor horsepower
The horsepower equivalent to the watts of electric power supplied to a motor.

wire-to-water efficiency
The ratio of the total power input (electric current expressed as motor horsepower) to a motor and pump assembly, to the total power output (water horsepower), expressed as a percent; the combined efficiency of the pump and motor.

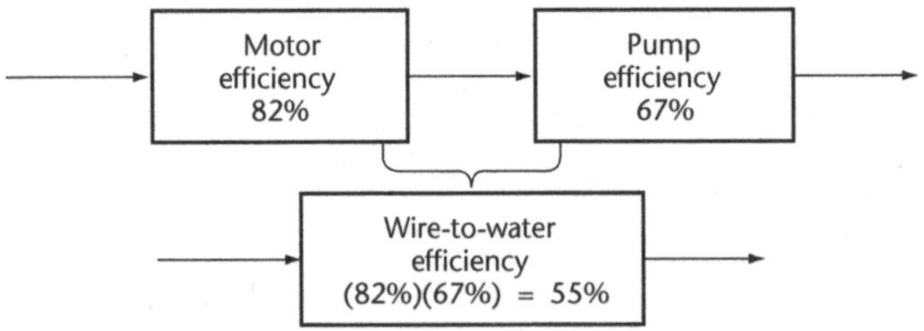

In practical horsepower calculations, you must take into account pump and motor efficiencies so that you can determine the size of the pump and motor and the amount of power necessary to move the water at the desired flow rate.

To show how horsepower and efficiency calculations are used, consider a system in which the pump is pumping 800 gpm against a total head of 46.2 ft. The water horsepower required to perform this work is calculated as follows:

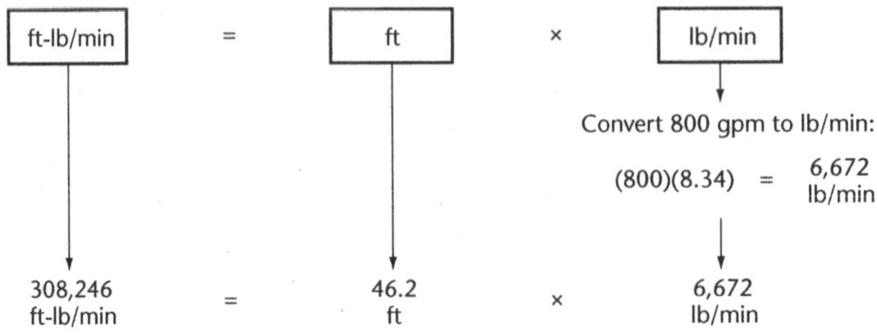

Then convert foot-pounds per minute to horsepower.

$$\frac{308{,}246 \text{ ft-lb/min}}{33{,}000 \text{ ft-lb/min/hp}} = 9.34 \text{ hp}$$

Now, suppose that the pump is only 85 percent efficient and the motor 95 percent efficient. To account for these inefficiencies, more than 9.34 whp would have to be supplied to the pump, and even more horsepower would have to be supplied to the motor.

Techniques for calculating the required horsepower are discussed in Example 7. To use the techniques, however, you must know that power input to the motor is usually expressed in terms of electrical power rather than horsepower.

mhp
or watts
or kilowatts
→ Motor

The relationship between horsepower and watts or kilowatts is given below. With these equations, you can convert from one term to the other.

1 horsepower = 746 watts power

1 horsepower = 0.746 kilowatts power

Example 7

If a pump is to deliver 460 gpm of water against a total head of 95 ft, and the pump has an efficiency of 75 percent, what horsepower must be supplied to the pump?

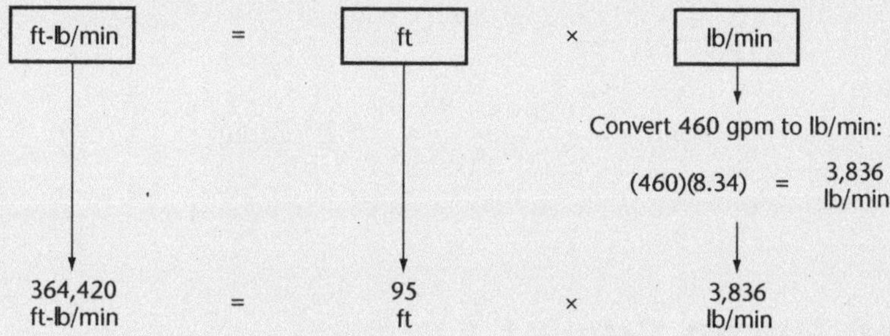

In this problem, brake horsepower is to be calculated. First calculate the water horsepower, based on the work to be accomplished. Then determine the brake horsepower required. Convert foot-pounds per minute to water horsepower:

$$\frac{364{,}420 \text{ ft-lb/min}}{33{,}000 \text{ ft-lb/min/hp}} = 11.04 \text{ whp}$$

This calculation shows that 11.04 whp (water horsepower) is required.

Now determine the amount of horsepower that must be supplied to the pump (brake horsepower):

? bhp → Pump → 11.04 whp

(75% efficient)

As stated in the problem, the pump is 75 percent efficient. Therefore, the brake horsepower will have to be more than 11.04 hp in order to accomplish the 11.04 hp work. Because the usable brake horsepower (75 percent) must equal 11.04 hp, the equation is

$$(75\%)(\text{bhp}) = 11.04 \text{ whp}$$

Restate the percentage as a decimal number:

$$(0.75)(x \text{ bhp}) = 11.04 \text{ whp}$$

And then solve for the unknown value:

$$(0.75)(x \text{ bhp}) = 11.04$$

$$x = \frac{11.04}{0.75}$$

$$x = 14.72 \text{ bhp}$$

Therefore, the 75 percent efficient pump must be supplied with 14.72 hp to accomplish the 11.04 hp of work:

14.72 bhp → Pump → 11.04 whp

(75% efficient)

Using the alternative equation, we simply need to insert the efficiency into the denominator, and so the equation becomes:

$$\text{hp} = \frac{\text{gpm} \times \text{head (ft)}}{3{,}960 \times \text{Eff.}}$$

Recalculating Example 7:

$$\text{hp} = \frac{460 \text{ gpm} \times 95 \text{ ft}}{3{,}960 \times 0.75} = 14.71 \text{ hp}$$

Reading Pump Curves

There is a great deal of information written about pumps, pump performance, and pump curves. The following section covers just those basic topics needed to understand and use pump curves.

Description of Pump Curves

pump characteristic curve

A curve or curves showing the interrelation of speed, dynamic head, capacity, brake horsepower, and efficiency of a pump.

A pump curve is a graph showing the *characteristics* of a particular pump. For this reason, these graphs are commonly called **pump characteristic curves**. The following characteristics are commonly shown on a pump curve:

- Capacity (flow rate)
- Total head
- Power (brake horsepower)
- Efficiency

If a pump is designed to be driven by a variable-speed motor, then "speed" is also a characteristic shown on the graph. For the purposes of this discussion, it will be assumed that the pump is driven by a constant-speed motor, so the graphs used will have only four curves.

The four pump characteristics (capacity, head, power, and efficiency) are related to each other. This is extremely important, for it is this interrelationship that enables the four pump curves to be plotted on the same graph.

Experience has shown that the *capacity* (flow rate) of a pump changes as the head against which the pump is working changes. Pump capacity also changes as the *power* supplied to the pump changes. Finally, pump capacity changes as *efficiency* changes. Consequently, head, power, and efficiency can all be graphed "as a function of" pump capacity. This simply means that *capacity* is shown along the horizontal (bottom) scale of the graph, and *head, power*, and *efficiency* (any one or a combination of them) are shown along the vertical (side) scales of the graph. The following material discusses each of these relationships individually.

The H–Q Curve

The *H–Q* curve shows the relationship between total head, *H*, against which the pump must operate and pump capacity, *Q*. A typical *H–Q* curve is shown in Figure 2-25. The curve indicates what flow rate the pump will produce at any given total head.

For example, if the total head is 100 ft, the pump will produce a flow of about 1,910 gpm. The *H–Q* curve also indicates that the capacity of the pump *decreases* as the total head *increases*. (Generally speaking, when the force against which the pump must work increases, the flow rate decreases.) The way total head controls the capacity is a characteristic of a particular pump.

The *H–Q* curve also identifies two very important operational facts. Normally on the curve you will find a mark like this: ¬. The mark defines the design point, the head and capacity at which the pump is intended to operate for best efficiency in a particular installation.

The P–Q Curve

The *P–Q* curve (Figure 2-26) shows the relationship between power, *P*, and capacity, *Q*. In this figure, pump capacity is measured as gallons per minute, and power is measured as brake horsepower. For example, if you pump at a rate of 1,480 gpm, then the power used to drive the pump is about 58 bhp. This is valuable information. As explained in the preceding section on horsepower

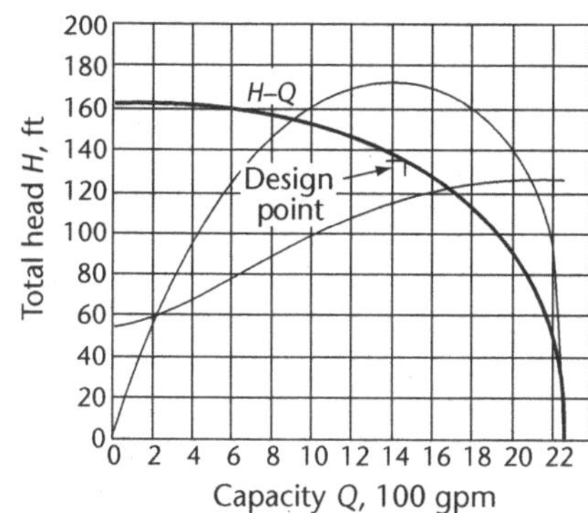

Figure 2-25 *H–Q* curve

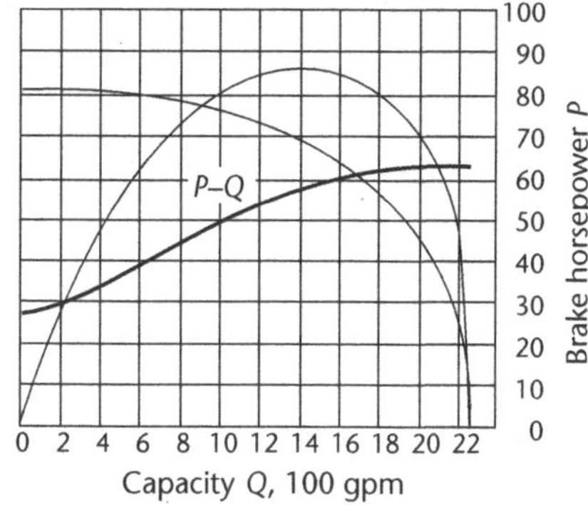

Figure 2-26 *P–Q* curve

design point
The mark on the H-Q (head-capacity) curve of a pump characteristics curve that indicates the head and capacity at which the pump is intended to operate for best efficiency in a particular installation.

calculations, if you know the brake horsepower and the motor efficiency, you can then determine the motor horsepower (the power required to drive the motor).

Knowledge of what power the pump requires is valuable for checking the adequacy of an existing pump and motor system. The knowledge is also important when you find yourself scrambling to rig a temporary system from spare or surplus parts.

The E–Q Curve

The *E–Q* curve shows the relationship between pump efficiency, *E*, and capacity, *Q*, as shown in Figure 2-27. (Sometimes the Greek letter eta—η—is used to represent efficiency, and the efficiency–capacity curve is designated the η–*Q* curve.) In sizing a pump system, the engineer attempts to select a pump that will produce the desired flow rate at or near peak pump efficiency.

The more efficient your pump is, the less costly it is to operate. Stated another way, the more efficient your pump is, the more water you can pump for each dollar's worth of power. Knowledge of pump efficiency allows you to compute the cost of pumping water. Therefore, from an operational point of view, pump efficiency is a "must know" item.

Reading the Curve

Figure 2-28 is the complete pump curve. By using the pump curve, you can determine any three of the four pump characteristics (capacity, head, efficiency, and horsepower) when given information about the fourth characteristic. Normally the total head or the capacity is the known characteristic. The following example illustrates how to read pump curves.

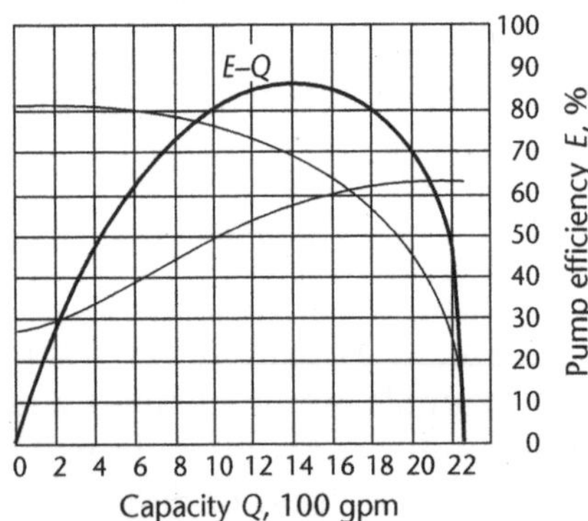

Figure 2-27 *E–Q* curve

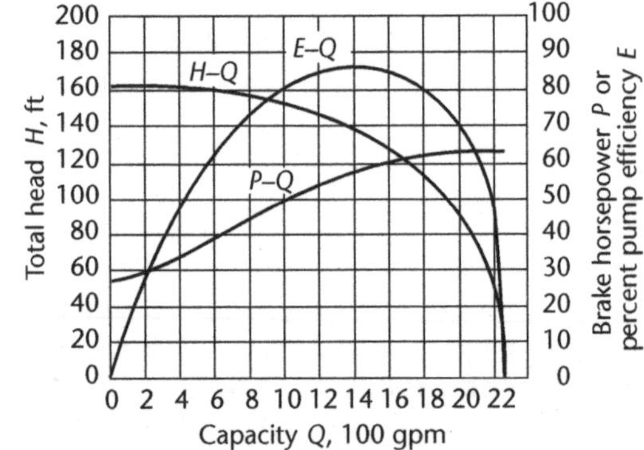

Figure 2-28 Complete pump curve

Example 8

Using the pump curve given in Figure 2-28, determine the pump head, power, and efficiency when operating at a flow rate of 1,600 gpm.

The key to reading pump curves is to find a certain vertical line and to note where the line intersects all three pump characteristic curves. In this problem, the known characteristic is *capacity* (1,600 gpm). To determine head, power, and efficiency, you should first draw a vertical line upward from the number 16 on the horizontal scale (Figure 2-29).

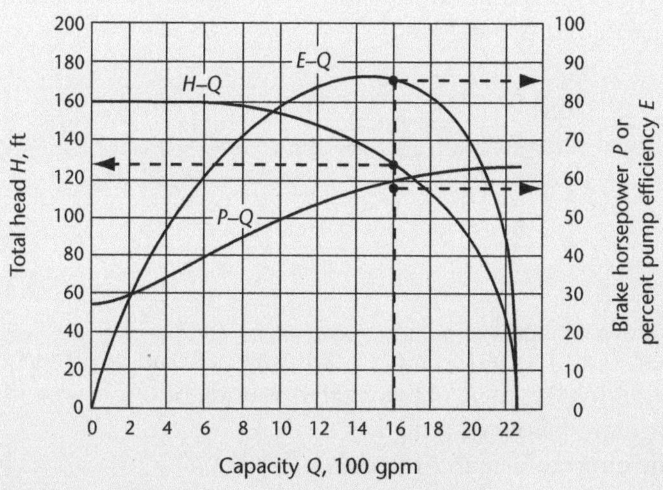

Figure 2-29 Pump curve for Example 8

Note that the vertical line intersects each of the three curves, as indicated by the three dots. Following the line upward, you first reach the intersection with the *P–Q* (power) curve. To determine what the power is at this point, move *horizontally to the right* toward the bhp scale; you should hit the scale at about 60 bhp.

Now return to the vertical line (1,600 gpm) and move to the intersection with the *H–Q* (head) curve. To determine the head at this point, move *horizontally to the left*, toward the total head scale; you should hit the scale at about 128 ft.

Finally, look at the point where the vertical line (1,600 gpm) intersects the *E–Q* (efficiency) curve. To determine the efficiency at this point, move *horizontally to the right* toward the efficiency scale. The reading is about 85 percent.

From the known pump capacity of 1,600 gpm, then, the following characteristics were determined:

- Brake horsepower P = 60 bhp
- Pump head H = 128 ft
- Pump efficiency E = 85%

WATCH THE VIDEO

Pump Curves (www.awwa.org/wsovideoclips)

Calculations for Chlorine Dosage, Demand, Residual, and Contact Time

The chlorine requirement, or chlorine dosage, is the sum of the chlorine demand and the desired chlorine residual. This can be expressed mathematically as follows:

$$\text{chlorine dosage (mg/L)} = \text{chlorine demand (mg/L)} + \text{chlorine residual (mg/L)}$$

This formula is often used in conjunction with the formula for converting milligrams per liter to pounds per day:

$$\text{feed rate (lb/d)} = \text{dosage (mg/L)} + \text{flow rate (mgd)} \times \text{conversion factor (8.34 lb/gal)}$$

Example 9

A water is tested and found to have a chlorine demand of 6 mg/L. The desired chlorine residual is 0.2 mg/L. How many pounds of chlorine will be required daily to chlorinate a flow of 8 mgd?

Apply the chorine demand formula:

chlorine dosage (mg/L) = chlorine demand (mg/L) + chlorine residual (mg/L)

Fill in the information given in the problem:

dosage = 6 mg/L + 0.2 mg/L

= 6.2 mg/L

Now use Equation 2-2 to convert the dosage in milligrams per liter to a feed rate in pounds per day:

feed rate = (dosage)(flow rate)(conversion factor)

= (6.2)(8)(8.34)

= 413.66 lb/d

Example 10

The chlorine demand of a water is 5.5 mg/L. A chlorine residual of 0.3 mg/L is desired. How many pounds of chlorine will be required daily for a flow of 28 mgd?

chlorine dosage (mg/L) = chlorine demand (mg/L) + chlorine residual (mg/L)

dosage = 5.5 mg/L + 0.3 mg/L

= 5.8 mg/L

Convert dosage in milligrams per liter to feed rate in pounds per day:

(5.8)(28)(8.34) = 1,354 lb/d chlorine feed rate

Note that if prechlorination is performed, it is not usually necessary to achieve a residual. Setting the residual equal to zero produces the following equivalencies:

chlorine dosage (mg/L) = chlorine demand (mg/L) + chlorine residual (mg/L)

chlorine dosage (mg/L) = chlorine demand (mg/L) + 0

chlorine dosage (mg/L) = chlorine demand (mg/L)

Therefore, when no residual is required, as in most prechlorination, the dosage is equal to the chlorine demand of the water.

The equation used to calculate chlorine dosage can also be used to calculate chlorine demand when dosage and residual are given, or to calculate chlorine residual when dosage and demand are given. The following examples illustrate the calculations.

Example 11

The chlorine dosage for a water is 5 mg/L. The chlorine residual after 30 min contact time is 0.6 mg/L. What is the chlorine demand in milligrams per liter?

$$\text{chlorine dosage (mg/L)} = \text{chlorine demand (mg/L)} + \text{chlorine residual (mg/L)}$$

$$5 \text{ mg/L} = x \text{ mg/L} + 0.6 \text{ mg/L}$$

Now solve for the unknown value:

$$5 \text{ mg/L} - 0.6 \text{ mg/L} = x \text{ mg/L}$$

$$4.4 \text{ mg/L} = x$$

Example 12

The chlorine dosage of a water is 7.5 mg/L, and the chlorine demand is 7.1 mg/L. What is the chlorine residual?

$$\text{chlorine dosage (mg/L)} = \text{chlorine demand (mg/L)} + \text{chlorine residual (mg/L)}$$

Fill in the given information and solve for the unknown value:

$$7.5 \text{ mg/L} = 7.1 \text{ mg/L} + x \text{ mg/L}$$

$$7.5 \text{ mg/L} - 7.1 \text{ mg/L} = x \text{ mg/L}$$

$$0.4 \text{ mg/L} = x$$

Sometimes the dosage is not given in milligrams per liter, but the feed rate setting of the chlorinator and the daily flow rate through the plant are known. In such cases, feed rate and flow rate should be used to calculate the dosage in milligrams per liter. The dosage in milligrams per liter can then be used in Equation 2-1, as in Examples 11 and 12.

$C \times T$ value
The product of the residual disinfectant concentration *C*, in milligrams per liter, and the corresponding disinfectant contact time *T*, in minutes. Minimum $C \times T$ values are specified by the Surface Water Treatment Rule as a means of ensuring adequate kill or inactivation of pathogenic microorganisms in water.

Example 13

Adequate chlorine (or other disinfectant) contact time is a requirement for surface water treatment plants and is called **$C \times T$ value**. It determines disinfection sufficiency and is the product of the disinfectant residual concentration (C) and the effective contact time (*T*) for the chamber or clearwell in which disinfection takes place. To calculate $C \times T$ values, operators need to know the chlorine residual in the chamber, the flow rate of water through the chamber, the baffling factor (T10) of the chamber, and the volume of the chamber. The results are shown in mg-min/L.

A disinfection chamber with a chlorine residual of 1.6 mg/L measures 44 ft by 30 ft and has a baffling factor of 0.34. What is the $C \times T$ value at 300 gpm if the chamber holds 8.3 ft of water?

$$\text{Effective volume} = \text{L} \times \text{W} \times \text{H} \times \text{baffling factor}$$

$$\text{Effective volume of chamber} = 44 \text{ ft} \times 30 \text{ ft} \times 8.3 \text{ ft} \times 7.48 \text{ gal/ft}^3 \times 0.34$$

$$= 27{,}863.3 \text{ gallons}$$

$$\text{Effective contact time (ECT)} = 27{,}863.3 \text{ gal/300 gpm} = 92.9 \text{ min}$$

$$C \times T \text{ value} = \text{ECT} \times \text{residual} = 92.9 \text{ min} \times 1.6 \text{ mg/L}$$

$$= 148.6 \text{ mg-min/L}$$

The baffling factor of a chamber is always less than 1 and is an indicator of the short-circuiting characteristics encountered in the chamber. Chambers with larger baffling factors are preferred because they yield higher $C \times T$ values.

WATCH THE VIDEO
Chlorination Breakpoint (www.awwa.org/wsovideoclips)

Study Questions

1. What is plotted on the horizontal scale (x-axis) of a pump curve?
   a. Efficiency
   b. Capacity (flow rate)
   c. Total head
   d. Power

2. What is the motor horsepower (mhp), if 200 horsepower (hp) is required to run a pump with a motor efficiency of 88% and a pump efficiency of 74%? Note: The 200 hp in this problem is called the water horsepower (whp). The whp is the actual energy (horsepower) available to pump water. Give results to two significant figures.
   a. 130 mhp
   b. 180 mhp
   c. 200 mhp
   d. 310 mhp

3. How many gallons per minute should a flowmeter register if a 10.0-in. diameter main is to be flushed at 5.10 ft /sec?
   a. 1,050 gpm
   b. 1,100 gpm
   c. 1,250 gpm
   d. 1,350 gpm

4. A tank 84.0 ft in diameter and 24.25 ft high at the overflow requires disinfection. How much 12.5% sodium hypochlorite that is 9.59 lb/gal will be required for a dosage of 50.0 mg/L?
   a. 310 gal
   b. 350 gal
   c. 380 gal
   d. 410 gal

5. A well that is 210 ft in depth and 14.0 in. in diameter requires disinfection. The depth to water from the top of the casing is 91 ft. If the desired dose is 50.0 mg/L, what is the number of pounds and ounces of sodium hypochlorite (12.5% available chlorine) required? Assume the sodium hypochlorite solution is 9.59 lb/gal.
   a. 43 oz of NaOCl
   b. 45 oz of NaOCl
   c. 49 oz of NaOCl
   d. 54 oz of NaOCl

6. What is the net positive suction head available (NPSHA) given the following data? Will the pump cavitate if the net positive suction head required (NPSHR) is 18.4 ft? Note: There are 1.11 ft/in. of Hg.

   - Atmospheric pressure (AP) = 29.8 in. Hg
   - Static suction lift (SSL) = 15.1 ft
   - Friction head loss (Hf) = 0.61 ft
   - Vapor pressure at 12°F (VP) = 0.50 ft

   a. 14 ft, therefore NPSHA < NPSHR so cavitation should occur
   b. 17 ft, therefore NPSHA < NPSHR so cavitation should occur
   c. 20 ft, therefore NPSHA > NPSHR so cavitation should not occur
   d. 22 ft, therefore NPSHA > NPSHR so cavitation should not occur

7. A storage tank has a capacity of 34.0 ft. Currently there are 22.89 ft of water in the tank. What would the SCADA reading be on the board in milliamps (mA) for a 4-mA to 20-mA signal?
   a. 13.9 mA
   b. 14.1 mA
   c. 14.3 mA
   d. 14.8 mA

8. The _____ is the outlet, or high-pressure, side of a pump.
   a. suction side
   b. flow gauge
   c. piezometer
   d. discharge side

9. What is the term for the amount of energy used by water in moving from one location to another?
10. What is the formula for calculating chlorine demand?
11. How is the wire-to-water efficiency obtained?
12. What is the formula to convert milligrams-per-liter concentration to pounds per day?

Chapter 3

Water Use and System Design

Many considerations are involved in planning and designing a water distribution system. Some of the factors that may affect the design are the source or sources of water, population density, economic conditions of the community, geographic location, and history and practices of the water system.

Water Source Effects on System Design

The type and location of the water source have considerable effect on the design, construction, and operation of a water distribution system. Systems can be classified, generally speaking, by the type of water source:

- Surface water
- Groundwater
- Purchased water
- Rural water

Surface Water Systems

It is rare for groundwater to be available in large enough quantities to support a large community, so many medium-size and essentially all large water systems use surface water sources (Figure 3-1). One of the prime features of a surface water system is that the water often enters from one side of the distribution system. As a result, large-diameter transmission mains are usually required to carry water to the far sides of the distribution system.

Some exceptions exist, but in general, surface water is of good quality and plentiful. This in turn attracts industries that require process water for cooling, cleaning, and incorporation into a product. The availability of good-quality water at a reasonable price generally promotes rapid growth of the community, which in turn causes frequent expansion of the water distribution system.

Groundwater Systems

Although groundwater is generally available in most areas of the United States, the amount available for withdrawal at most locations is limited. If groundwater is generally available and the water requires no special treatment, some water systems are able to install several wells at various locations in the distribution system. Groundwater systems may require few if any transmission mains because water flows from several directions through the piping grid.

surface water system
A water system using water from a lake or stream for its supply.

groundwater system
A water system using wells, springs, or infiltration galleries as its source of supply.

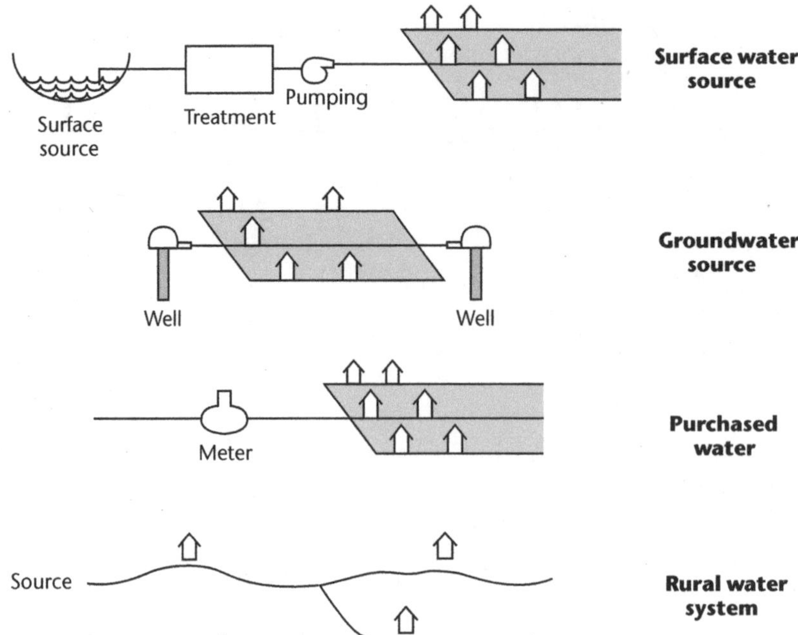

Figure 3-1 Water systems classified by source

If groundwater is available at only one location, all of the wells may be located at one side of the distribution system, so the piping design is similar to that of a surface water system. A similar situation arises if the groundwater must receive treatment for contaminant removal or aesthetic improvement. In this case, water from several wells is piped to a central treatment facility and then pumped to the distribution system.

Purchased Water Systems

Many small water systems that started out using water from their own wells eventually had to change to either a surface water source or purchased water when their community outgrew the capacity of the groundwater source. Other communities have switched to purchased water when it was discovered that their groundwater source was contaminated and treatment for contaminant removal was not economically practical.

Examples of large numbers of purchased water systems are in the Chicago, Illinois, and Detroit, Michigan, areas where hundreds of surrounding communities draw water from a few large treatment plants using water from the Great Lakes. Some of these systems rechlorinate the water as it enters their systems, but otherwise no treatment is necessary.

Purchased water systems must usually provide a large amount of water storage because they depend on a single connection. If the connection should break, they could be without water for hours, or even days. Purchased water systems must maintain particularly tight water accountability because they are paying for all water metered to them, including unmetered uses and water wasted in leaks.

purchased water system
A water system that purchases water from another water system and so generally provides only distribution and minimal treatment.

rural water system
A water system that has been established to serve widely spaced homes and communities in areas having no available groundwater or having water of very poor quality.

Rural Water Systems

Another class of water system has developed in recent years that has distribution system design and operating problems somewhat different than those of other water utilities. Rural water systems have developed in areas where both groundwater and surface water are nonexistent or of extremely poor quality. Many rural systems have been funded by government programs.

The systems obtain water from remote sources, treat it if necessary, and then run long mains across the countryside to provide water to individual farms, homes, and small communities.

In most cases, the water mains are plastic pipes installed by plowing them into the ground, and, in most systems, there is no intent to provide fire protection. The water main capacity is sufficient only to provide domestic water in limited quantities. Operators of these types of rural water systems face many unique problems in operating and maintaining their systems.

Types of Water System Layout

The three general ways in which distribution systems are laid out include an arterial-loop system, a grid system, and a tree system.

Arterial-loop systems (Figure 3-2) are designed to have large-diameter mains around the water service area. Flow will be good at any point within the grid, because water can be supplied from four directions.

Grid systems (Figure 3-3) have most of the water mains that serve homes and businesses interconnected, and they are reinforced with larger arterial mains that feed water to the area. If the grid mains are all at least 6 in. (150 mm) in diameter, flow is usually good at most locations because water can be drawn from two or three directions.

Tree systems (Figure 3-4) have transmission mains that supply water into an area, but the distribution mains that branch off are generally not connected, and many are dead ends. This is usually considered poor design because flow to many locations is through only one pipeline. Flow near the end of a long "branch" may be relatively poor. This design is poor also because a relatively large number of customers may be without water while repairs are made at a point near the connection of the branch to the transmission main. Customers at the ends of long branches may complain of poor water quality owing to the poor circulation of water in the system.

Unfortunately, few distribution systems are completely laid out in an ideal pattern. Most systems have been added onto as new housing or industrial areas have

arterial-loop system
A distribution system layout involving a complete loop of arterial mains (sometimes called trunk mains or feeders) around the area being served, with branch mains projecting inward. Such a system minimizes dead ends.

grid system
A distribution system layout in which all ends of the mains are connected to eliminate dead ends.

tree system
A distribution system layout that centers around a single arterial main, which decreases in size with length. Branches are taken off at right angles, with subbranches from each branch.

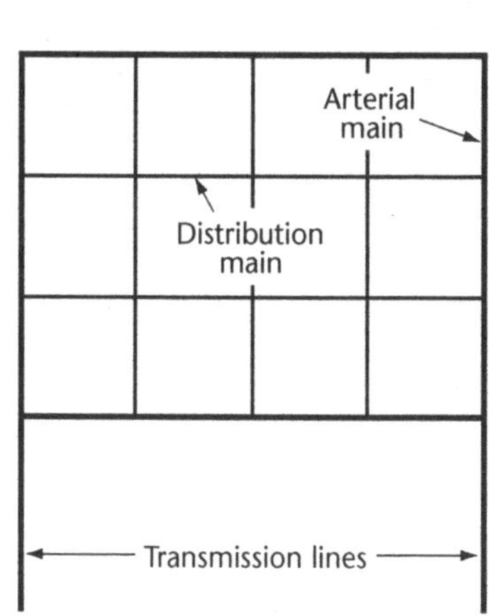

Figure 3-2 Arterial-loop system

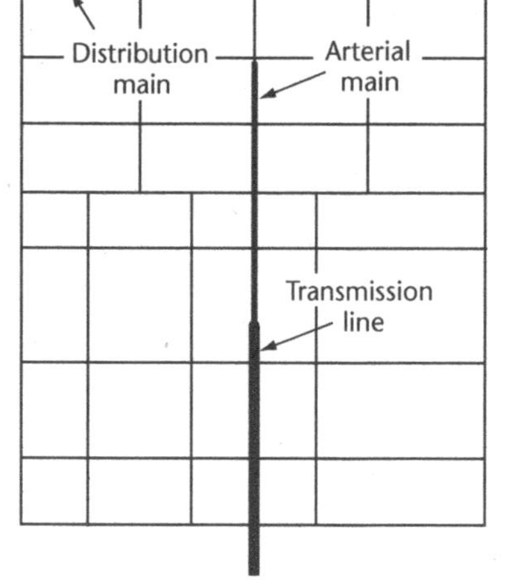

Figure 3-3 Grid system

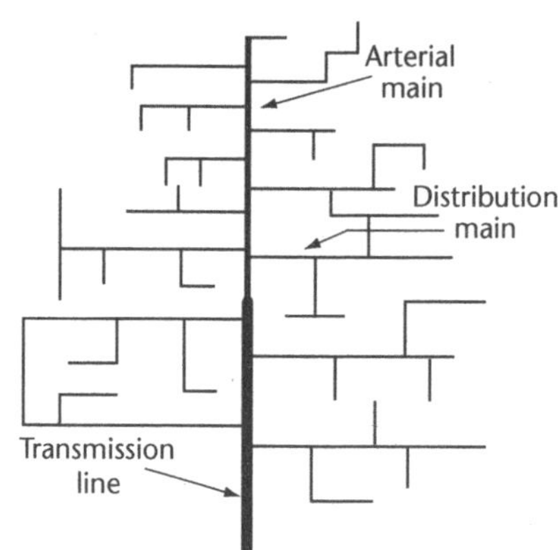

Figure 3-4 Tree system

been developed and the original systems required reinforcement to carry the additional loads. Most distribution systems actually combine grid and tree systems.

Dead-End Mains

Residents usually like to live on dead-end streets because there is very little traffic.

From a water supply standpoint, though, dead-end streets are undesirable because it is difficult to "loop" the piping system. A dead-end main can cause the following types of problems:

- The fire hydrants on the dead-end main draw from only one direction, so they may not provide adequate flow.
- Domestic use on the main provides a very low flow rate, so water quality in the dead-end main often degrades to the point of prompting customer complaints about taste, odor, or rusty water. Many water systems have had to set up regular schedules of flushing dead-end mains to avoid these customer complaints.

If a dead-end main is at least 6 in. (150 mm) in diameter and has sufficient flow and pressure, a fire hydrant should be installed at the end of the pipe. If the main does not have sufficient capacity to provide minimum fire flow, an approved flushing hydrant or blowoff should be installed for flushing purposes. Flushing devices should be sized to provide a flow velocity of at least 2.5 ft/sec (0.76 m/sec) in the water main.

Valves and Hydrants

Shutoff valves should be installed at frequent intervals in the distribution system so that areas may be isolated for repair without having to shut off too many customers. As a rule, at least two valves should be installed at each main intersection, and three is much better. Where there are long sections of main with no intersections, intermediate valves should also be installed at what would normally be one-block intervals. The Ten States Standards (promulgated in 2012 by the Great Lakes–Upper Mississippi River Board of State Public Health and Environmental Managers) recommend a maximum spacing of 500 ft (152 m) in business districts, 800 ft (244 m) in residential areas, and 1 mi (1.6 km) in rural areas where future development is not expected.

Unless there are special circumstances, the best location for valves at a street intersection is opposite the right-of-way line for the intersecting street. This usually keeps them beyond the street paving and makes them easy to find. Mid-block valves should be located opposite an extended lot line, which is least likely to place them in a residential or business driveway.

Fire hydrants are best located at street intersections, and if the blocks are long, additional hydrants should be located near the middle of each block. Hydrant spacing should generally be between 350 and 600 ft (107 and 183 m), depending on the density and valuation of the area being served.

Water Main Sizing

Water mains must carry water for many reasons, including the following:

- Domestic use in residential homes and apartments
- Irrigation of plants and lawns
- Commercial uses, such as those at stores, public buildings, and industries
- Fire flow

Water Use Terms

The following terms are frequently used in determining pipe sizing for distribution system design:

- **Average day demand** is the total system water use for 1 year divided by 365 days in a year.
- **Per-capita water use** is the average day demand divided by the number of residents connected to the water system. This figure varies widely from system to system, depending primarily on the quantity of water used for commercial and irrigation purposes. Nationally, the figure is estimated to be about 105 gpcd (gallons per capita per day) (397 L/d per capita).
- **Maximum day demand** is the water use during the 24 hours of highest demand during the year.
- **Peak hour demand** is the water use during the highest 1-hour period during the year or sometimes during the history of the system.

Residential and Commercial Water Use

Residential flow is very small in comparison to the amount of water that can be carried in water mains of the size required for fire flow. In areas where there is no restriction on water use for sprinkling, the maximum hourly water use on a hot summer day may be two or three times the average rate, but this is still small in comparison to the capacity of mains required for fire flow.

Some industries and businesses, such as hospitals, incorporate large quantities of water in their products. Except under extreme emergencies, normal water service must be maintained to key facilities, such as hospitals, at the same time as fire flow is maintained in the area. In this case, the average water use of key facilities must be added to the fire flow requirements when designing mains.

Fire Flow Requirements

The primary factor in determining the size of water mains, storage tanks, and pumping stations for water systems serving a population of fewer than 50,000 is usually the fire protection requirements, commonly referred to as **fire flow** requirements. The requirements for each community are set by the Insurance Services Office (ISO), which represents the fire insurance underwriters in the United States. The ISO determines the minimum flow that the water system must be able to maintain for a specified period of time in order to receive a specified fire protection rating. The fire insurance rates in the community are then based, in part, on this classification.

Many small, older water systems were originally built with 4-in. (100-mm) diameter and smaller mains. These small lines will often provide adequate capacity for domestic flow, but fire hydrants located on them will not yield adequate flow to meet ISO requirements. The general requirements for a well-designed system meeting fire flow requirements are as follows:

- Mains in residential areas should have a minimum size of 6 or 8 in. (150 or 200 mm).
- Mains serving business, industrial, or other high-value areas should be at least 8 in. (200 mm) and may need to be larger to provide adequate fire flow.
- Mains smaller than 6 in. (150 mm) should be installed only when they are to be used to provide circulation in a grid system. The general policy today is that water mains smaller than 6 in. (150 mm) should not have fire hydrants connected to them.

average day demand
The total system water use for 1 year divided by 365 days in a year.

per-capita water use
The average day demand divided by the number of residents connected to the water system.

maximum day demand
The water use during the 24 hours of highest demand during the year.

peak hour demand
The greatest volume of water in an hour that must be supplied by a water system during any particular time period, such as a year, to meet customer demand.

fire flow
The rate of flow, usually measured in gallons per minute (gpm) or liters per minute (L/min), that can be delivered from a water distribution system at a specified residual pressure for firefighting. When delivery is to fire department pumpers, the specified residual pressure is generally 20 psi (140 kPa).

Additional information on fire flow requirements can be found in AWWA Manual M31, *Distribution System Requirements for Fire Protection.*

Water Pressure Requirements

The normal working pressure in the distribution system should preferably be between 35 and 65 psi (241 and 448 kPa). Higher pressures will significantly increase main and service leaks and will hasten the failure of water heaters and water-using appliances. In addition, customers do not like very high pressure because it will blow dishes out of their hands if faucets are opened quickly. The Uniform Plumbing Code requires that water pressure not exceed 80 psi (552 kPa) at service connections, unless the service is provided with a pressure-reducing valve.

The minimum pressure at ground level, at all points in the distribution system, *under all flow conditions*, must be 20 psi (138 kPa). In other words, the pressure should not drop below this pressure during fire flow conditions.

A water system supplying an area that has varying elevation must usually be divided into pressure zones to maintain reasonable pressure to customers. Water is furnished initially to the highest zone and then admitted to the lower zones through pressure-reducing valves.

Water Velocity Limitations

The velocity of water flow in pipes is also a consideration in determining pipe sizes. The limit for normal operations should be about 5 ft/sec (1.5 m/sec). When the velocity is higher, friction loss becomes excessive. Higher velocities can usually be tolerated under fire flow conditions.

Sizing to Maintain Water Quality

When sizing water mains, there is a temptation to add a "safety" factor, enlarging the mains an additional size. This practice should be avoided because long detention times may result, which degrades water quality. The minimum size that can satisfy the fire and domestic demand should be selected.

Network Analysis

The sizing of water mains depends on a combination of factors, including system pressure, flow velocity, head loss resulting from friction, and the size of all mains that lead to a particular location. The calculation used for many years for analyzing flow in a distribution system involves use of the Hazen–Williams formula.

When a distribution system is to be expanded, it is usually necessary to analyze the entire system and to determine whether and how the system must be modified to properly handle the new loading. This analysis is often performed by computer modeling techniques.

Hydraulic computer models are an important tool used by most systems to help provide information as a basis for decisions regarding planning, design, and operation of the distribution network. The models use extensive and complex calculations to construct computer simulations of system performance under various operating conditions, thus providing insight for managers, engineers, and operators regarding the best options for their use. Distribution system network computer models serve many functions:

- Planning applications, including the following:
  - Capital improvement program
  - Conservation impact studies
  - Water main rehabilitation programs
  - Reservoir siting

- Engineer design applications, including the following:
  - Fire flow studies
  - Valve sizing
  - Reservoir sizing
  - Pump station location and sizing
  - Calculation of pressure and flow at various locations
  - Pressure zone boundary locating
- Operating applications, including the following:
  - Personnel training
  - Troubleshooting
  - Water loss calculations
  - Emergency operations simulations
  - Source management
  - Model calibration
  - Main flushing programs
  - Area isolation assistance
  - Energy cost control
- Water quality improvements, including the following:
  - Constituent tracking
  - Water source and water age tracking
  - Disinfectant residual level prediction
  - Water quality monitoring location selection

Distribution system models may use proprietary software or public domain programs like EPANET, which is provided by the US Environmental Protection Agency (USEPA). Most programs include a feature to export into the EPANET file format if desired. This USEPA program is available for public use. It is a program used mostly for network analysis and is the basis for most proprietary computer programs. It can perform steady-state and extended-period simulation of hydraulic and water quality in pipe networks.

Regardless of the computer network program used, it is critical to properly "calibrate" the model so that it will provide useful information that reproduces the operation of the physical distribution network.

Study Questions

1. The pressure during fire flow conditions should not drop below
   a. 15 psi (103 kPa).
   b. 20 psi (138 kPa).
   c. 25 psi (172 kPa).
   d. 30 psi (207 kPa).

2. Which type of distribution system configuration has smaller mains that generally terminate as dead ends?
   a. Grid system
   b. Dendritic system
   c. Arterial-loop system
   d. Tree system

3. How many times higher than the normal operating pressure should the pressure rating of distribution system piping be?
   a. 1.0–2.0 times
   b. 2.5–4.0 times
   c. 4.0–5.0 times
   d. 5.0–7.0 times

4. Regarding fire flow, mains smaller than 6 in. (150 mm) should be used only
   a. in residential areas.
   b. in low-value districts.
   c. in rural areas.
   d. to complete a grid.

5. What is the commonly accepted water velocity in water mains at maximum flows?
   a. 1–3 ft/sec (0.3–0.9 m/sec)
   b. 2–4 ft/sec (0.6–1.2 m/sec)
   c. 4–5 ft/sec (1.2–1.5 m/sec)
   d. 5–6 ft/sec (1.5–1.8 m/sec)

6. The size and capacity of a water distribution system is based largely on
   a. peak hour customer demand.
   b. fire demand.
   c. agricultural demand.
   d. industrial demand.

7. A(n) _____ layout centers around a single arterial main, which decreases in size with length.
   a. arterial-loop system
   b. tree system
   c. rural system
   d. grid system

8. What type of water system uses wells, springs, or infiltration galleries as its source of supply?

9. What type of water system is established to serve widely spaced homes and communities in areas having no available groundwater or having water of very poor quality?

10. In what type of water distribution layout are all ends of the mains connected to eliminate dead ends?

11. What is the term for the water use during the 24 hours of highest demand during the year?

Fluids at Rest and in Motion

Hydraulics is the study of fluids in motion or under pressure. An understanding of hydraulics is necessary for the proper operation of a water distribution system. In this book, the subject is confined to the behavior of water in water distribution systems. Some of the basic concepts used in the operation of water distribution systems are covered in this chapter.

Static Pressure

Water flows in a water system when it is under a force that makes it move. The force on a unit area of water is termed pressure. The pressure in a water system is a measure of the height to which water theoretically will rise in an imaginary standpipe open to atmospheric pressure. Static pressure exists although water does not flow; dynamic pressure exists as "moving energy."

All objects have weight because they are acted on by gravity. When a 1-lb brick is placed on a table with an area of 1 square inch (1 in.^{2}), it exerts a force of 1 pound per square inch (psi) on the table. Two stacked bricks on the 1-in.^{2} table would exert a force of 2 psi. But if the size of the brick is doubled to 2 in.^{2}, the pressure is halved. And if the size of the brick is tripled, the pressure in pounds per square inch is reduced by one-third.

Likewise, a column of water 10 ft high exerts a total force of 4.33 psi. If you connect a pressure gauge at the bottom of a water tube with 10 ft of water in it, the gauge will read 4.33 psi. If you connect the pressure gauge to the bottom of a larger-diameter column with 10 ft of water in it, the gauge will still read 4.33 psi (Figure 4-1). Water pressure is dependent only on the height of the column. However, the total weight exerted on the floor by the water in the large column will obviously be much more.

Dynamic Pressure

If the water in the column is permitted to empty horizontally from the bottom of a column, the water will begin to flow under the hydrostatic pressure applied by the height of the column. The flowing water will have little hydrostatic pressure, but it will have gained moving, dynamic pressure, or kinetic energy. The hydrostatic pressure is static potential energy converted into moving energy.

One can add energy to a water system and thereby increase hydrostatic and dynamic pressure. A pump does this when it pumps water into elevated storage. The hydrostatic pressure (height) to which the water can be pumped is equivalent to pressure (less losses) at the pump discharge.

hydraulics
The study of fluids in motion or under pressure.

pressure
The force on a unit area of water.

static pressure
Pressure that exists in water although the water does not flow.

dynamic pressure
Pressure that exists in water as moving energy.

hydrostatic pressure
The pressure exerted by water at rest (for example, in a nonflowing pipeline).

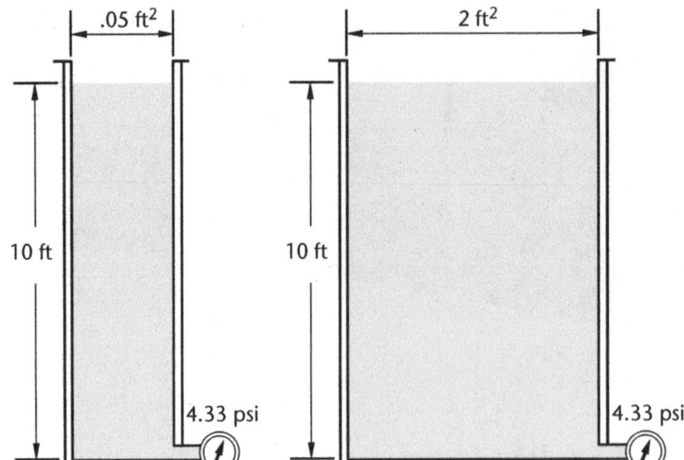

Figure 4-1 Hydraulic head depends only on column height

Pressure is usually measured in either pounds per square inch or feet of head in US units, or as kilopascals (kPa) of pressure or meters (m) of head in metric units. A pressure of 1 psi is equal to approximately 6.895 kPa.

Velocity

The speed at which water moves is called **velocity**, usually abbreviated *V*. The velocity of water is usually measured in feet per second (ft/sec) in US units and meters per second (m/sec) in metric terms. For comparison, a rapidly moving river might move at about 7 ft/sec (2.13 m/sec).

The quantity of water (*Q*) that flows through a pipe depends on the velocity (*V*) of the flow and the cross-sectional area (*A*) of the pipe. This relationship is stated mathematically as the formula $Q = A \times V$. Or, in terms of velocity,

$$V = \frac{Q}{A}$$

For example, a flume is 2 ft wide and 2 ft deep, so the cross-sectional area of the flume is 4 ft^2. The flume is flowing full of water and the quantity is measured at 12 ft^3 in 1 second (12 ft^3/sec). The velocity of the water would therefore be calculated as follows:

$$V = \frac{Q}{A}$$

$$= \frac{12 \text{ ft}^3\text{/sec}}{4 \text{ ft}^2} = 3 \text{ ft/sec}$$

Friction Loss

As water flows through a pipeline, there is friction between the water and the walls of the pipe. The friction loss causes a loss of head (pressure) as the water flows through the pipe. The amount of friction depends partly on the smoothness of the pipe walls. All new pipe is quite smooth, whereas old, badly corroded cast-iron pipe will have a very high friction factor. The degree of pipe roughness is commonly denoted by a *C* factor, which is a coefficient in the Hazen–Williams formula that has long been used for determining flow in pipe. High *C* values imply less friction.

velocity
The speed at which water moves; measured in ft/sec or m/sec.

The head loss due to friction also depends on the velocity of the water flowing through the pipe, the diameter of the pipe, and the distance the water travels through the pipe.

Figure 4-2 is a commonly used nomograph for approximating the flow in ductile-iron pipe. In the example shown by a dashed line, a 12-in. pipe is flowing at approximately 600 gpm and the pipe has a C factor of 140. A line is drawn from the 600-gpm point on the discharge line, through the point for 12-in. pipe, and to the pivot line. A line is then drawn from that point to 140 on the flow coefficient line. This line crosses the loss of head line at about 0.7, indicating this is the head loss per 1,000 ft of pipeline. If, for example, you are determining the loss of head in a pipeline 3,000 ft long with no valves or bends, the theoretical loss of head would be three times the indicated value (3 × 0.7 = 2.1 ft of head loss).

Pipe fittings also add a significant pressure loss in flow, and this is usually expressed as the equivalent length of straight pipe. To use the nomograph in Figure 4-3, a line is drawn from the pipe size to the point for each type of fitting, and the equivalent pipe length is read from the center scale. The total of all readings is then added to the actual length of the pipeline in determining the expected loss of head.

Referring to the dashed line in Figure 4-2, each medium sweep elbow in the previous 12-in. pipeline example would add friction loss equal to about 26 ft of pipe. So if the example pipeline has 20 elbows along the 3,000-ft length, it would

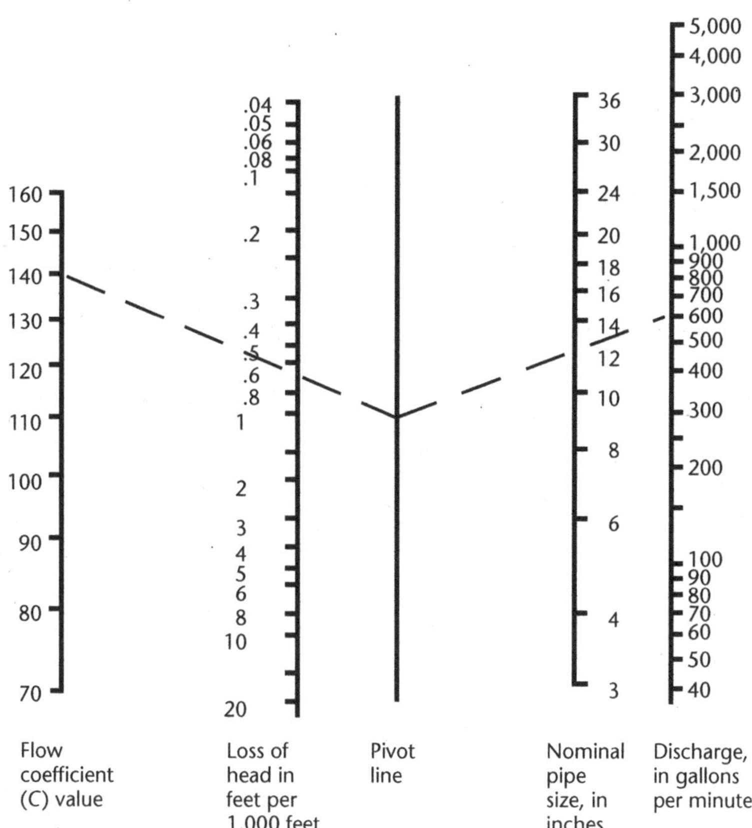

Figure 4-2 Flow of water in ductile-iron pipe

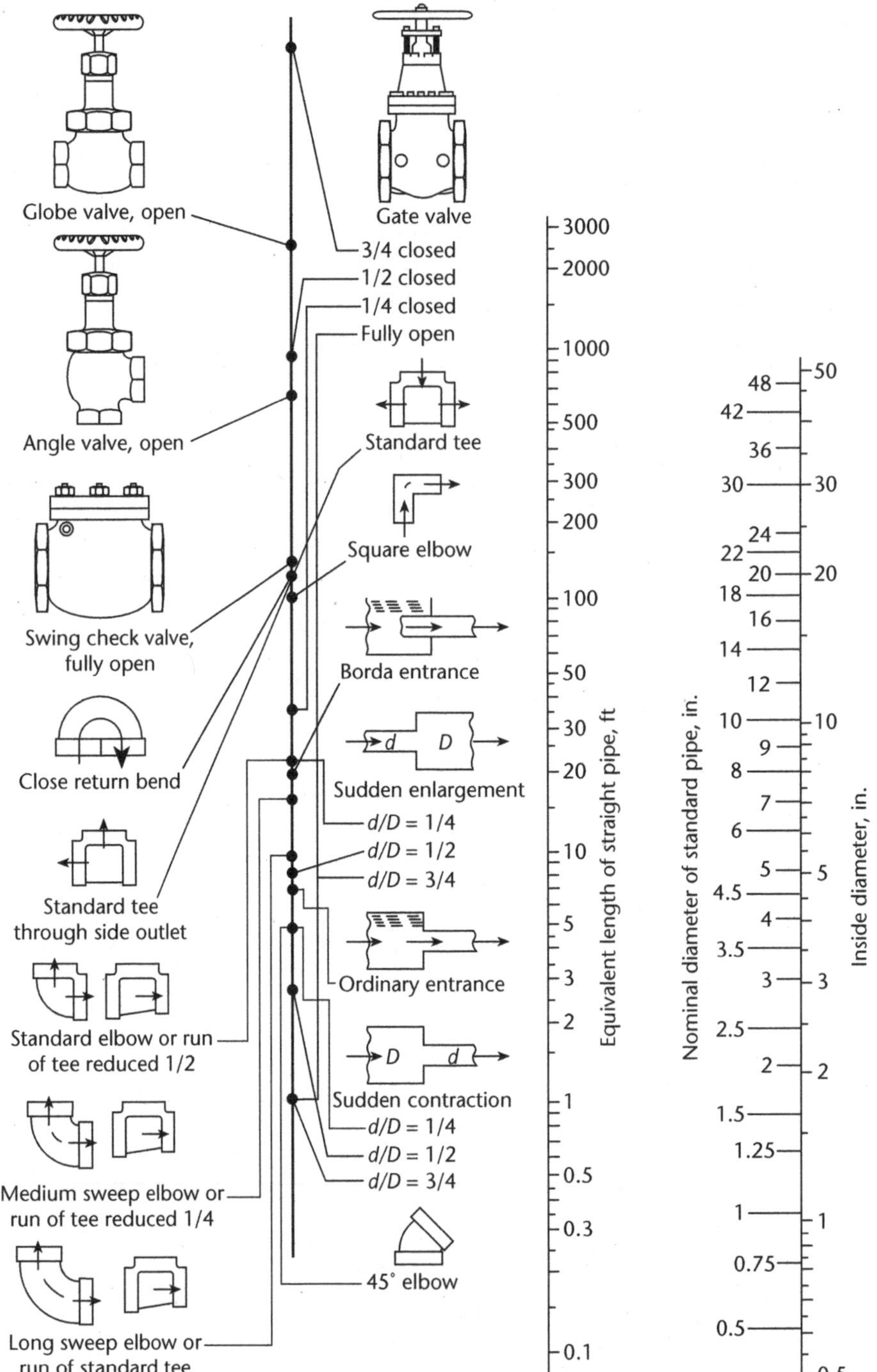

Figure 4-3 Resistance of valves and fittings to flow of fluids

add friction loss equal to an additional 20 × 26 = 520 ft of pipe. This friction loss would cause additional loss of head as follows:

$$\text{loss of head per 1,000 ft} = 0.7 \text{ ft}$$

$$\text{loss of head for 520 ft} = \frac{520}{1{,}000} \times 0.7 = 0.36 \text{ ft loss of head due to elbows}$$

This loss would be added to the loss of head determined for 3,000 ft of pipe, so the total loss would be 2.1 + 0.36 = 2.46 ft of head loss. If there are also tees, valves, and other fittings in the pipeline, the head loss that they cause can be computed and added to the total.

Table 4-1 Designation of US pipe sizes to the metric system

Customary Inches	Proposed Millimeters	Customary Inches	Proposed Millimeters
¼	8	16	400
⅓	10	18	450
½	15	20	500
¾	20	21	525
1	25	24	600
1¼	32	27	675
1½	40	30	750
2	50	33	825
2½	65	36	900
3	80	42	1,050
3½	90	48	1,200
4	100	54	1,350
6	150	60	1,500
8	200	66	1,650
10	250	72	1,800
12	300	78	1,950
14	350	84	2,100
15	375		

This example also illustrates that the loss in head can become quite significant over a long pipeline. If, for example, after adding up the losses caused by all the other fittings, the total loss of head in the pipeline is 5 ft, this loss in terms of pressure would be 5 ft × 0.433 lb/in.^{2}/ft = 2.17 psi. In other words, if the pressure entering the pipe is 50 psi, the theoretical pressure at the far end would be reduced to 50 – 21.65 = 28.35 psi.

To convert the information on the nomographs for metric use, refer to Table 4-1 for the metric equivalents of US unit pipe sizes. Flow in gallons per minute (gpm) can be converted to liters per second (L/s) when multiplying by 0.06308. Head of water expressed in feet can be converted to meters of water when multiplying by 0.3048.

Hydraulic Gradient

As discussed in Chapter 2, the head of water at any point in a water system refers to the height to which water would rise in a freely vented standpipe. The head at each point would be the height of the water column. The imaginary line joining the elevations of these heads is called the *hydraulic grade line*. The slope or steepness of this line is called the *hydraulic gradient*.

A simple hydraulic gradient is illustrated in Figure 4-4. Assuming there is equal flow in all sections of the line, the gradient becomes steeper as the pipe becomes smaller because of the friction head loss. If there were no flow in the line, the water head at the end of the line would be at the same level as the water in the reservoir.

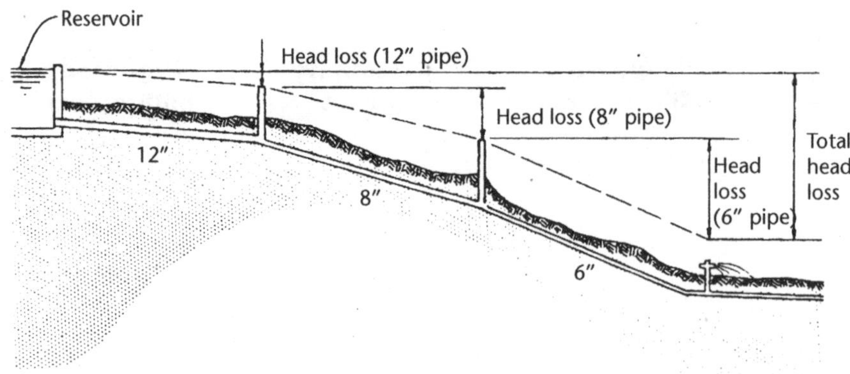

Figure 4-4 Pipe size affects hydraulic gradient

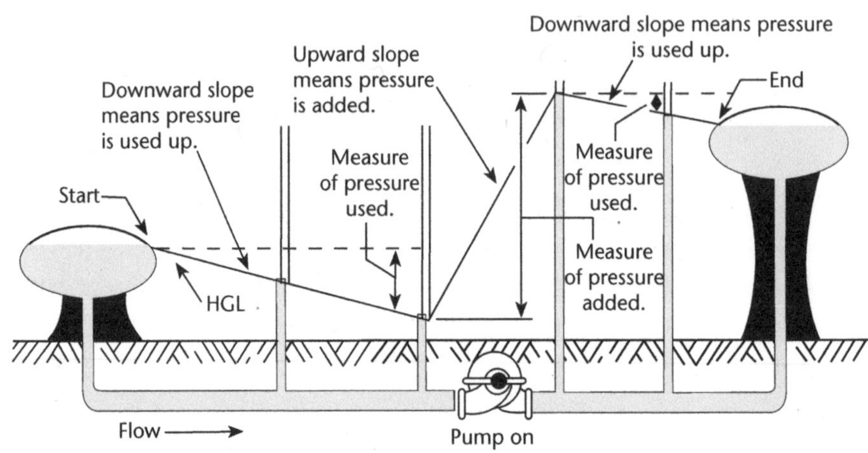

Figure 4-5 Five basic principles of hydraulic grade lines (HGLs) in dynamic systems

In the example in Figure 4-5, water flows from a reservoir toward a booster pump. The pump adds pressure and causes an upward slope of the hydraulic grade line, then the head falls as the water travels to the end of the system.

water hammer
The potentially damaging slam, bang, or shudder that occurs in a pipe when a sudden change in water velocity (usually as a result of someone too-rapidly starting a pump or operating a valve) creates a great increase in water pressure.

surge pressure
A momentary increase of water pressure in a pipeline due to a sudden change in water velocity or direction of flow.

Hydraulic Transients

Very brief but often drastic changes in flow within pipes can cause magnified pressure changes. These often short-lived (transient) events can result in broken pipelines, can damage pumping facilities, and may allow contaminants to enter the potable water system. Hydraulic transients can cause either high or low pressures and may last from milliseconds to several seconds.

Water Hammer

The most common hydraulic transient encountered in water distribution systems is water hammer, also called surge pressure. Water hammer occurs most often when a valve is closed too quickly. This causes a pressure wave to develop within the pipe. The magnitude of the maximum pressure is related to the velocity of the water, the pressure at the time, and the rate at which the valve was closed. Water hammer has caused rupture of pipe and pump casings; pipe collapse; vibration; excessive pipe displacement; pipeline fittings failure; and cavitation.

System operators should take care to avoid creating water hammer. Always operate valves slowly to eliminate water hammer. Educate anyone who may use fire hydrants to open and close with care.

Surge Control

A specialized type of water hammer can be caused by sudden stoppage of system pumps. This stoppage often occurs during a power outage. When the pump stops, water from the system (perhaps from a filling storage reservoir) can surge back toward the pump. The resulting water hammer can damage pumps, pipelines, fittings, and other equipment. Surge control equipment should be installed on all distribution system pumps (see Figures 4-6 through 4-11). Typical surge control

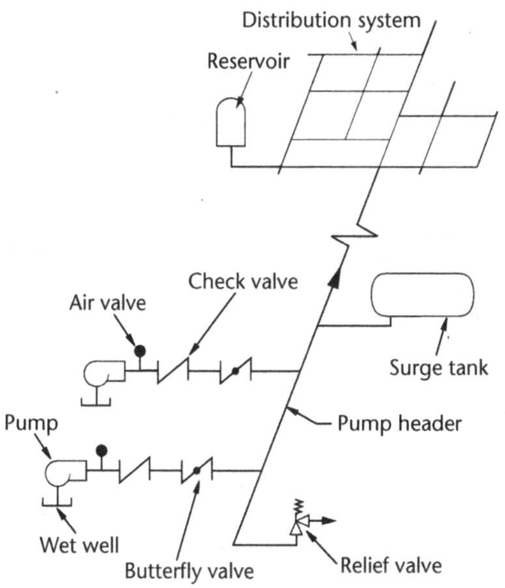

Figure 4-6 Typical pumping/distribution system

Courtesy of Wal-Matic Valve and Manufacturing Corporation.

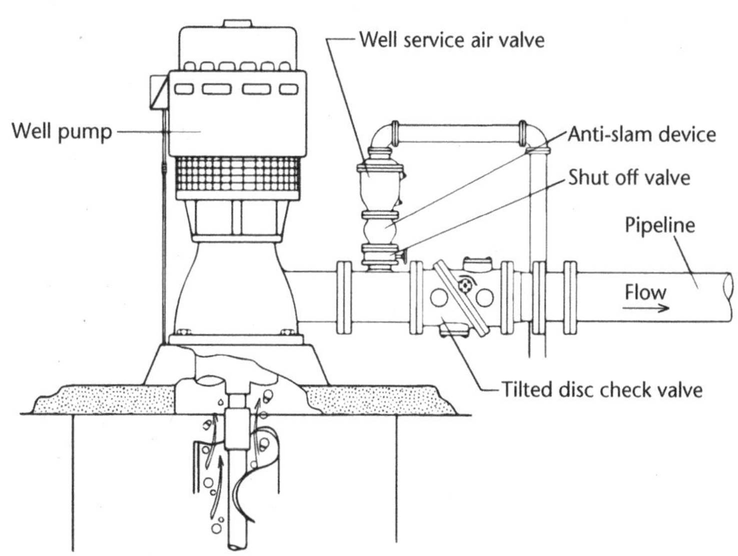

Figure 4-7 Vertical turbine well pump with a well service air valve

Courtesy of Wal-Matic Valve and Manufacturing Corporation.

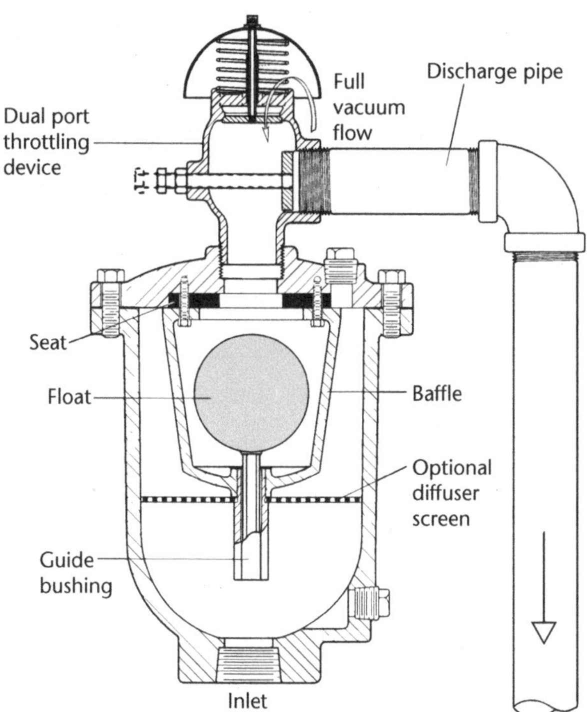

Figure 4-8 Well service air valve

Courtesy of Wal-Matic Valve and Manufacturing Corporation.

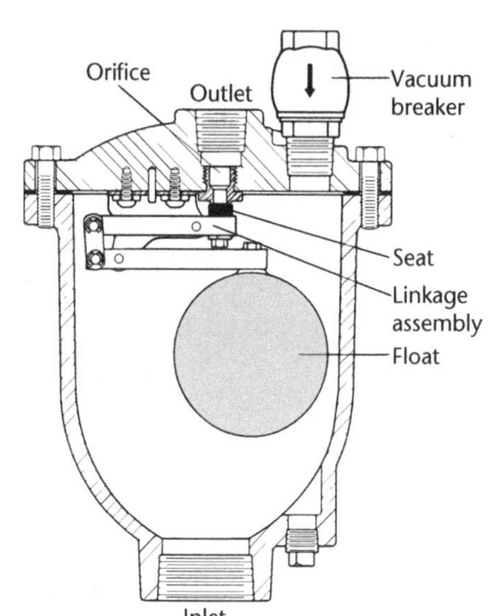

Figure 4-9 Air-release valve

Courtesy of Wal-Matic Valve and Manufacturing Corporation.

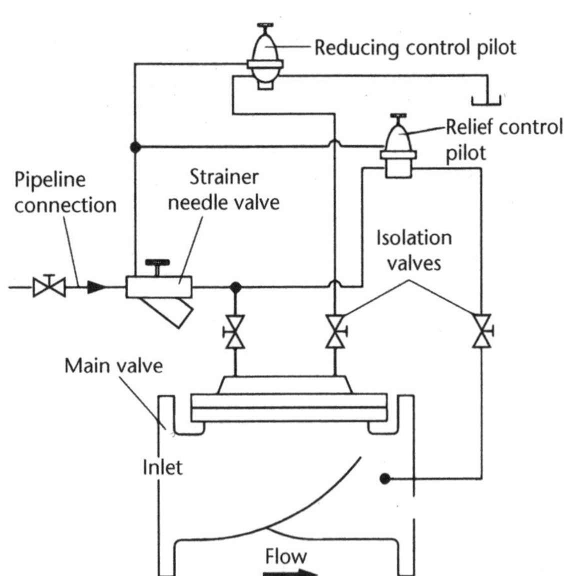

Figure 4-10 Surge relief valve and anticipator valve

Courtesy of Wal-Matic Valve and Manufacturing Corporation.

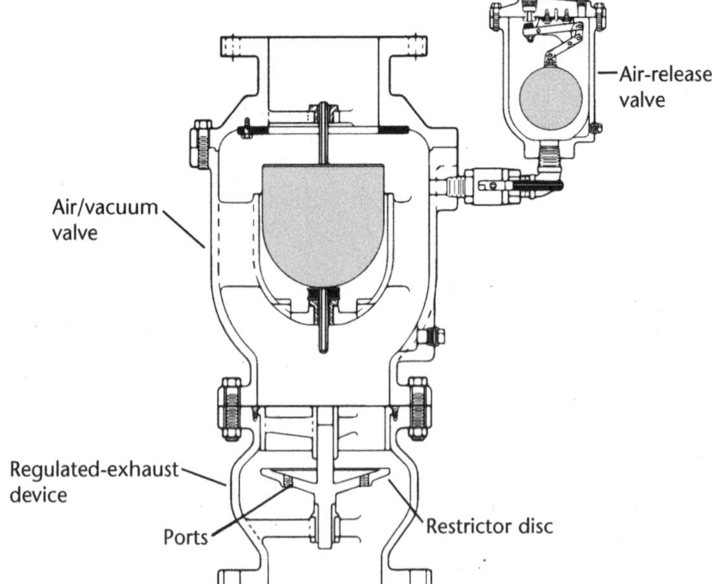

Figure 4-11 Surge-suppression air valve

Courtesy of Wal-Matic Valve and Manufacturing Corporation.

systems include pump control valves, surge tanks, air valves, relief valves, and check valves. These devices require regular maintenance according to manufacturer recommendations.

Study Questions

1. As the total head on a system increases, the volume that a centrifugal pump delivers is reduced
   a. directly.
   b. indirectly.
   c. proportionately.
   d. disproportionately.

2. Which type of distribution system configuration has interconnected mains?
   a. Grid system
   b. Dendritic system
   c. Arterial-loop system
   d. Tree system

3. Water behind a dam and above a water treatment plant has energy by virtue of its elevation. This difference in elevation is called elevation head or
   a. kinetic energy.
   b. velocity head.
   c. potential energy.
   d. pressure head.

4. Although velocity head can usually be ignored, it should be considered when it is ____ to _____ percent of the pressure head.
   a. 1; 2
   b. 3; 4
   c. 5; 8
   d. 5; 10

5. Hydraulics is the study of
   a. fluid pressure in pipes or conduits.
   b. the force of fluids in motion.
   c. the pressure of fluids in motion.
   d. fluids in motion and at rest.

6. Which of the following is the most common cause for surges in the distribution system?
   a. Power failure shutting down a pump suddenly
   b. Valve opening and closing
   c. Failure of flow on pressure regulation
   d. Pump startup

7. _____ is the speed at which water moves and is measured in ft/sec or m/sec.
   a. Pressure
   b. Head
   c. Friction
   d. Velocity

8. System operators should take care to avoid creating _____ by always operating valves slowly.
   a. head loss
   b. water hammer
   c. friction
   d. sedimentation

9. What type of pressure exists in water although the water does not flow?

10. When converting from US pipe sizes, what is considered the metric equivalent of 0.25 in.?

11. What is the formula for calculating the quantity of water that flows through a pipe?

Chapter 5

Instrumentation and Control

Secondary Instrumentation and Telemetering

Secondary instrumentation displays the signals from the sensors of primary instrumentation. It also allows distant control. Secondary instruments are usually panel mounted. They can be mounted in local control panels, filter control consoles, area control panels, or main control panels.

Before World War II, most measuring and control instruments were mounted adjacent to the process being controlled, with direct connections to the process. These primary measuring devices required operators to move throughout the plant to take readings and make control adjustments. With a small process plant and an abundant workforce, this was sufficient.

Then, as plant monitoring, maintenance, and compliance requirements increased, monitoring and control functions needed to be more centralized. Direct connection was no longer possible, so secondary instrumentation was developed. Secondary instrumentation measures the parameters and transmits signals that correspond to the measurements. Secondary instrumentation required the development of a signal transmission method, as well as field and panel-mounted hardware to perform the monitoring and control functions.

Pneumatic and Electronic Signal Transmission

The first signal transmission methods used air pressure (i.e., pneumatic) transmission. Electronic signal transmission, developed later, uses electrical current or voltage signals. Both methods perform the same functions, and the instrument exteriors look more or less identical, both in the field and in the control room. As development and usage have progressed, both systems have also standardized signal levels. The pneumatic standard range is 3–15 psig (20–100 kPa [gauge]), whereas the electronic standard is 4–20 mA DC (direct current). The practical advantage of this standardization is that control room instruments' operating mechanisms are based on common signal units in all locations. That is, the panel may contain indicators, recorders, and controllers handling pressure, temperatures, flows, and other process variables, but the input and output signals operate over the same standard range. The only differences among them are the display scales or charts employed.

Pneumatic and electronic signal transmission systems can also be used within the same process control system, allowing a great deal of flexibility in providing instrumentation for a plant. Equipment can be chosen that most effectively suits the application and environment. Signal converters are readily available to convert pneumatic pressure to electric current (P/I converters) and electric current to pneumatic pressure (I/P converters).

secondary instrumentation
Instruments that respond to and display information from primary instrumentation.

pneumatic
Operated by air pressure.

Although pneumatic systems are still in use, most new systems are electronic. The costs of maintenance and installation of pneumatic systems have made them virtually obsolete. The remainder of this chapter describes the components of electrical systems.

Receivers and Indicators

receiver
(1) The part of a meter that converts the signal from the sensor into a form that can be read by the operator; also called the *receiver–indicator.*
(2) In a telemetry system, the device that converts the signal from the transmission channel into a form that the indicator can respond to.

indicator
The part of an instrument that displays information about a system being monitored. Generally either an analog or digital display.

analog
Continuously variable, as applied to signals, instruments, or controls.

digital
Varying in precise steps, as applied to signals or instrumentation and control devices.

Receivers convert the signal sent by the transmitter to an indicator reading for the operator to monitor. The **indicator** may be

- a direct-reading display that shows the current value of the parameter being monitored,
- a recorder that preserves the information for later examination,
- a totalizer that gives the total accumulated value since the instrument was last reset, or
- some combination of these units.

Indicator displays and recorders are of two types: analog and digital. These terms are commonly used in describing various components and functions of instrumentation. **Analog** values range smoothly from the minimum to the maximum value of a given range. An analog signal is either a variable voltage or current. The dial indicators shown in Figures 5-1 and 5-2 are examples of analog displays. Analog indicators include dial gauges and strip or circle charts (Figures 5-3 and 5-4). The indicated values on an analog display range smoothly from the display's zero to its maximum.

Digital values, on the other hand, take on only a fixed number of values within a range. Digital indicators, like digital watches, display decimal numbers. The number of possible readings within a given range is limited by the number of digits displayed.

Most parameters measured in water distribution are continuous in nature, like an analog signal or display. However, analog-to-digital converters allow continuous values to be displayed on digital indicators or transmitted over digital transmission channels. And digital-to-analog converters allow the reverse conversion.

Digital indicators are usually more accurate than analog indicators. They are not subject to the errors associated with electromechanical or mechanical systems, and they are easier to read correctly. However, analog indicators may be preferable for at-a-glance monitoring to ensure a value remains within a given range or to observe its rate of change.

Figure 5-1 Analog indicator

Figure 5-2 Analog and digital indicator

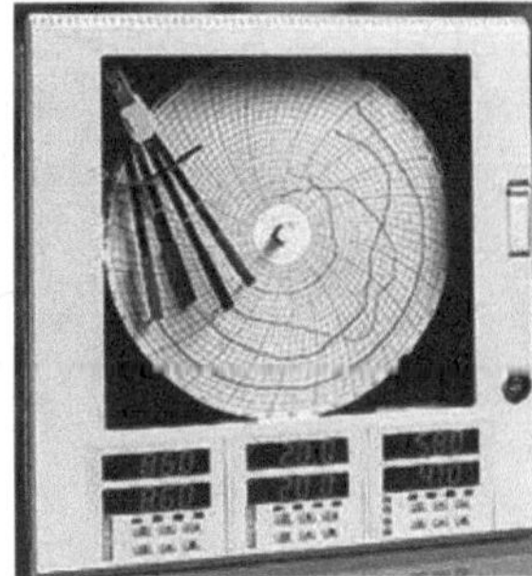

Figure 5-3 Circular recorder

Figure 5-4 Strip chart recorder

Telemetry

To monitor conditions at very distant locations, such as a remote pump station or reservoir, a telemetry system may be used. With this system, a sensor is connected to a transmitter, which sends a signal over a transmission channel to a combination receiver–indicator.

The type of signal used must be designed to maintain its accuracy over a long distance. Older equipment used audio tones or electrical pulses, but most equipment now transmits the information by a digital signal. The signal is transmitted either through direct wiring, through a leased telephone line, or by radio or microwave transmission (Figure 5-5). The receiver converts the signal to operate the indicator.

Telemetry systems allow flow rate, pressure, and other distribution system parameters to be sensed at one or more remote sites and indicated at a central location. Every telemetry system has the following three basic components:

1. Transmitter
2. Transmission channel
3. Receiver

The transmitter takes in data from one or more sensors at the remote site. It converts the data to a signal that is sent to the receiver over the transmission channel. The receiver changes the signal into standard electric values that are used to drive indicators and displays, recorders, or automatic control systems.

Telemetry Transmission Channels

The transmission channel in a telemetry system may be cable owned by a water utility that extends for short distances, such as between two buildings on a common site. In most cases, however, the channel is either a leased telephone line, a radio channel, or a microwave system. A system using space satellites is also available, though it is expensive. The leased telephone line may be a dedicated metallic pair, which is relatively expensive but highly reliable and interference-free or it may be a standard voice-grade phone line. Most modern transmitters generate signals that are designed to be sent over voice-grade lines and fiber-optic cables. Cellular communications may also be used to transmit signals.

Radio channels can be in the VHF (very high frequency) or UHF (ultra high frequency) band. Both radio and microwave systems generally require a

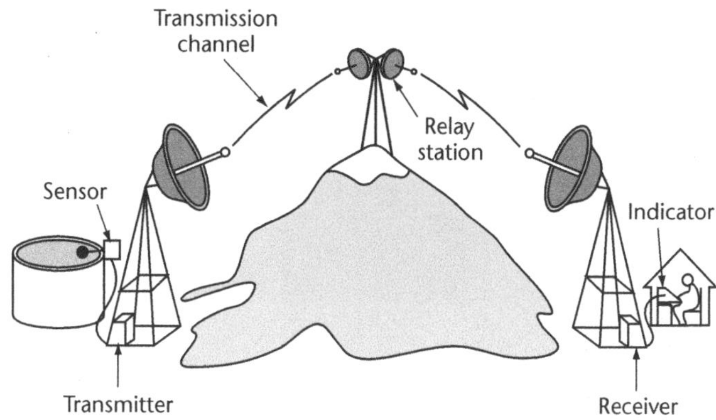

Figure 5-5 Relay station for radio or microwave telemetry system

telemetry
A system of sending data over long distances, consisting of a transmitter, a transmission channel (wire, radio, or microwave), and a receiver. Used for remote instrumentation and control.

receiver–indicator
An instrument component that combines the features of a receiver and an indicator.

transmitter
In telemetry or remote instrumentation, the device that converts the signal generated by the sensor into a signal that can be sent to the receiver–indicator over the transmission channel.

transmission channel
In a telemetry system, the wire, radio wave, fiber-optic line, or microwave beam that carries the data from the transmitter to the receiver.

current
(1) The flow rate of electricity, measured in amperes. (2) In telemetry, a signal whose amperage varies as the parameter being measured varies.

voltage
(1) A measure of electrical potential. (2) In telemetry, a type of signal in which the electromotive force varies as the parameter being measured varies.

pulse-duration modulation (PDM)
A telemetry-signaling protocol in which the time that a signal pulse remains on varies with the value of the parameter being measured.

variable frequency
Relating to a type of telemetry signal in which the frequency of the signal varies as the parameter being monitored varies.

line-of-sight path, which is unobstructed by buildings or hills between the transmitter and the receiver. To bypass obstructions or to ensure signal strength over very long distances, relay stations may be required.

Analog Signal Systems

Commonly used analog signals include the following:

- Current. The DC current generated by the transmitter is proportional to the measured parameter.
- Voltage. The DC voltage generated by the transmitter is proportional to the measured parameter.
- Pulse-duration modulation (PDM). The time period that a signal pulse is on is proportional to the value of the measured parameter.
- Variable frequency. The frequency of the signal varies with the measured parameter.

Current and voltage signals can be used only for short-distance systems with utility-owned cable or a leased metallic pair. The signals can be damaged by line loss and other factors over telephone lines. PDM and variable-frequency signals can be used for any distance over any type of channel.

Digital Signal Systems

Digital systems generally send binary code, in which the transmitter generates a series of on–off pulses that represent the exact numerical value of the measured parameter (e.g., off–on–off–on represents 5). These signals can be used over long or short distances with any transmission channel. The binary code signal is well adapted for connection to computerized systems. Figure 5-6 illustrates analog and digital telemetry signals.

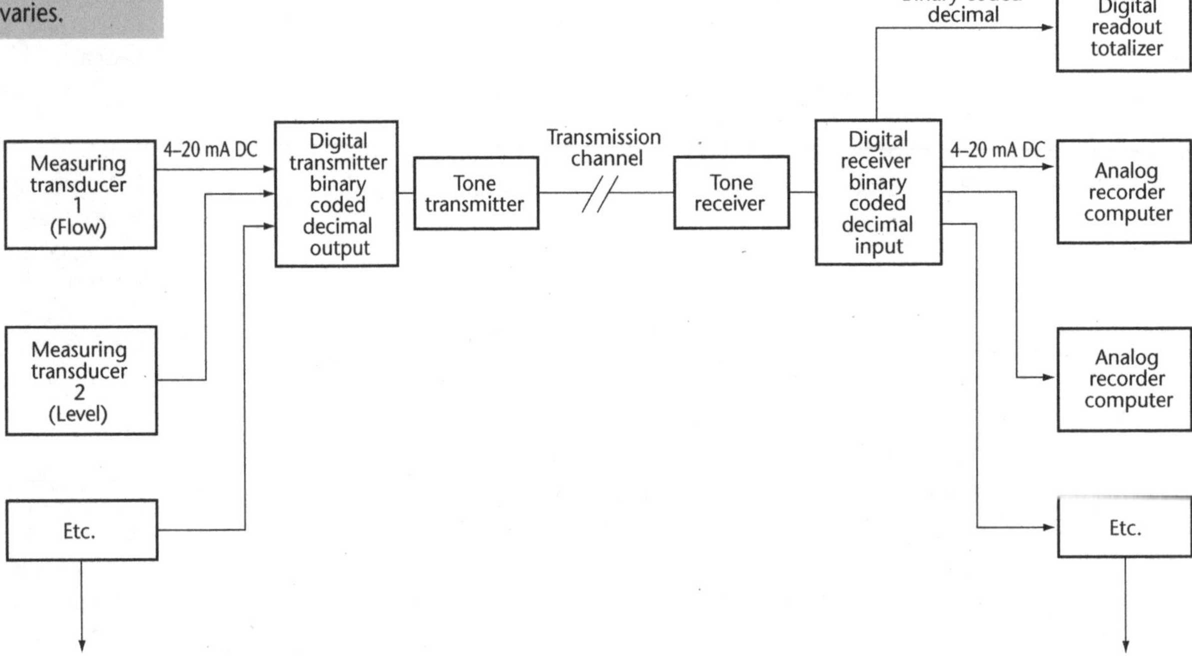

Figure 5-6 Typical digital telemetering system

In digital systems, the remote or transmit device is normally referred to as a remote terminal unit (RTU) and the receiver is known as the control terminal unit (CTU). A rather significant difference between analog and digital telemetry systems is that in digital systems, the RTU does not itself directly measure the variables in the system. Instead, the variables are measured by a transducer device, which normally converts the physical value of the process variable into a current output signal of 4–20 mA DC. These signals in turn serve as input to the RTU. The RTU converts the data into a message consisting of digital words that are transmitted to the CTU.

Multiplexing

Multiplexing systems allow a single physical channel, such as a single phone line or radio frequency, to carry several signals simultaneously. Tone-frequency multiplexing accomplishes this by having tone-frequency generators in the transmitter and tone-frequency filters in the receiver. The signal representing each measured parameter is assigned a separate audio frequency. The transmitter sends data representing each signal only over its assigned frequency, and the filters in the receiver allow it to respond to each frequency separately. Up to 21 distinct frequencies can be sent over a single voice-grade line. Tone-frequency multiplexing can be used with PDM and digital systems.

Scanning

Scanning is a second method of sending multiple signals over a single line or transmission channel. A scanner at the transmitter end checks and transmits the value of each of several parameters one at a time, in a set order. The receiver decodes the signal and displays each value in turn. Scanning can be used with all types of signals and with all types of transmission channels. Scanning and tone-frequency multiplexing can be combined to allow even more signals over a single line. A 4-signal scanner combined with a 21-channel tone frequency multiplexer would yield 84 distinct signal channels on a single line.

Polling

Another method of using a single line or channel to send several different signals is known as polling. In a polling system, each instrument has a unique address, or identifying number. A system controller unit sends messages over the line telling the instrument at a given address to transmit its data. The process of asking an instrument to send data is called polling.

In many systems, the controller is programmed to poll instruments as often as necessary to monitor the system; some instruments may need to be checked more often than others. In more sophisticated systems, the controller regularly scans the status of each instrument to see whether there is new information to be transmitted. If the status indicates that new information exists, the controller instructs the instrument to send its data. In some systems, critical instruments also have the capability to interrupt a long transmission by another instrument in order to call the controller's attention to urgent new data. Because data are transmitted only when needed under the polling system, a single line or channel can handle more instruments than with a simple scanning system.

remote terminal unit (RTU)
A computer terminal used to monitor the status of control elements, monitor and transmit inputs from instruments, and respond to data requests and commands from the master station.

control terminal unit (CTU)
The receiver in a digital signal system.

multiplexing
The use of a single wire or channel to carry the information for several instruments or controls.

scanning
A technique of checking the value of each of several instruments, one after another. Used to monitor more than one instrument over a single channel.

polling
A technique of monitoring several instruments over a single communications channel with a receiver that periodically asks each instrument to send current status.

Duplexing

In many telemetry installations, the instrument signals received at the operator's central location may require the operator to send control signals back to the remote site. Duplexing allows this to be accomplished with a single line:

1. Full duplex allows signals to pass in both directions simultaneously.
2. Half duplex allows signals to pass in both directions but only in one direction at a time.

Full-duplex systems usually employ tone-frequency generators to divide the line into transmission and receiving channels. Half-duplex systems may use tone-frequency generators, or they may simply rely on timing signals (similar to scanners) or on signals indicating status, such as end-of-transmission or ready-to-receive.

Control Systems

Control systems consist of three distinct components:

1. *Signal conditioners* receive either pneumatic, electric, or electronic signals from a controller. These signals are then conditioned or amplified and used to initiate the actuator. Signal conditioners include solenoids, starters, and positioners.
2. *Actuators* produce either rotary or linear movement of the final element. Actuators are usually motors or hydraulic cylinders and their related gearing (e.g., valve operators and motor controllers).
3. *Control elements* are equipment such as pumps and valves that change the process fluid.

Understanding the relationship among these three elements is important. In many cases, the elements are supplied by different manufacturers, and each component must meet the system requirements. The combination of these three will produce a final control element with its own distinct characteristics.

Two different types of controls are required in a process control system: two-state or continuous. Two-state control requires the final control element to be either on (open) or off (closed). Continuous control (also called modulation control) requires the element to vary its operation between the minimum and maximum points. An example is a valve operator, which may be designed to (1) operate a valve either fully opened or closed or (2) throttle the valve at intermediate positions.

In addition to the two types of control and the three components of final control elements, a variety of control media (air, electric, hydraulic) and several actuator types are available.

Control Classifications

Control equipment can be completely independent of instrumentation or it may operate in direct response to instrument signals. There are four principal classifications of control:

- Direct manual
- Remote manual
- Semiautomatic
- Automatic

duplexing
A type of telemetry in which a single line allows the operator to send the instrument signals that are received at the central location back to a remote site.

Direct Manual Control

In a **direct manual control** system, the operator directly operates switches or levers to turn the equipment on or off or otherwise change its operating condition. A valve operated by a handwheel is a common manually controlled piece of equipment. Operating electrical equipment requires throwing levers on the motor starter. Manual control has the advantages of low initial cost and no auxiliary equipment that must be maintained, but equipment operation may be time-consuming and laborious for the operator.

Remote Manual Control

With **remote manual control**, the operator is also required to turn a switch or push a button to operate equipment. However, the operator's controls may be located some distance from the equipment itself. When the operator activates the control switch, an electric relay, solenoid, or motor is energized, which in turn activates the equipment. Power valve operators and magnetic motor starters are common examples of remote manual control devices.

The solenoids and relays used for remote control are common components of all types of control systems. A **solenoid** (Figure 5-7) is an electric coil with a movable magnetic core. When an electric current is passed through the coil, a magnetic field is generated that pulls the core into the coil. The core can be attached to any piece of mechanical equipment that needs to be moved by remote control. Solenoid valves (Figure 5-8) are very common in water system controls.

A **relay** is constructed of a solenoid that operates an electrical switch or a bank of switches. The most common use of a relay is to allow a relatively low-voltage,

direct manual control
A type of system control in which personnel manually operate the switches and levers to control equipment from the physical location of the equipment.

remote manual control
A type of system control in which personnel in a central location manually operate the switches and levers to control equipment at a distant site.

solenoid
An electrical device that consists of a coil of wire wrapped around a movable iron core. When a current is passed through the coil, the core moves, activating mechanical levers or switches.

relay
An electrical device in which an input signal, usually of low power, is used to operate a switch that controls another circuit, often of higher power.

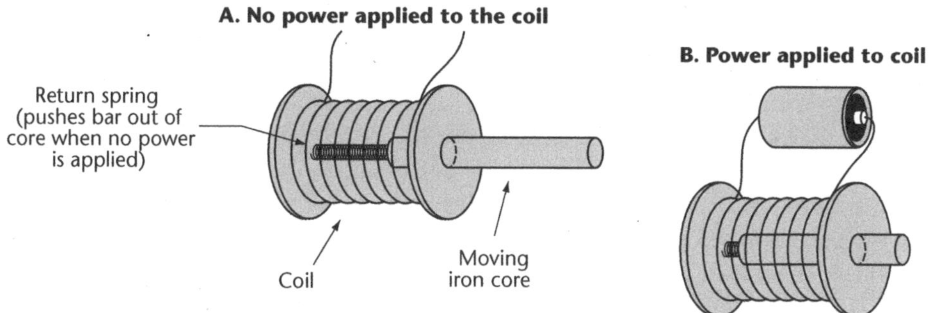

Figure 5-7 Operation of a solenoid

Figure 5-8 Solenoid valves
Courtesy of ASCO®.

low-current control circuit to activate a high-voltage, high-current power circuit. A typical power relay and a control relay are illustrated in Figure 5-9.

Semiautomatic Control

Semiautomatic control combines manual or remote manual control with automatic control functions for a single piece of equipment. A circuit breaker, for example, may disconnect automatically in response to an overload, then require manual reset.

Automatic Control

An automatic control system turns equipment on and off or adjusts its operating status in response to signals from instruments and sensors. The operator does not have to touch the controls under normal conditions. Automatic control systems are quite common. Simple examples are a thermostat used to control a heating system and automatic activation of lighting systems at night.

A number of modes (logic patterns) of operation (Figure 5-10) are available under automatic control. Two common modes are on–off differential control and proportional control.

1. On–off differential control is used to turn equipment full on when a sensor indicates a preset value, then turn it full off when the sensor indicates a second preset value.
2. Proportional control is used to open a valve or increase a motor's speed when the sensor shows a variation from a preset intended value. A common application is control of a chemical feeder in response to a flowmeter or residual analyzer signal.

Automatic controllers attempt to imitate human decision making, but they cannot achieve the level of complexity in decision making that humans can. Therefore, automatic control is limited to the more simple process situations. However, many water utility processes are in this category. The various pieces of equipment used to control each process parameter form a control loop of information processing. The direction of the information flow loop can be in either the same direction as the process (feedforward control) or the opposite direction of the process (feedback control).

A feedforward control loop measures one or more inputs of a process, calculates the required value of the other inputs, and then adjusts the other inputs to

semiautomatic control
A form of system control equipment in which many actions are taken automatically but some situations require human intervention.

automatic control
A system in which equipment is controlled entirely by machines or computers, without human intervention, under normal conditions.

on–off differential control
A mode of controlling equipment in which the equipment is turned fully on when a measured parameter reaches a preset value, then turned fully off when it returns to another preset value.

proportional control
A mode of automatic control in which a valve or motor is activated slightly to respond to small variations in the system, but activated at a greater rate to respond to larger variations.

Power relay

Control relay

Figure 5-9 Typical relays
Courtesy of Danaher Controls.

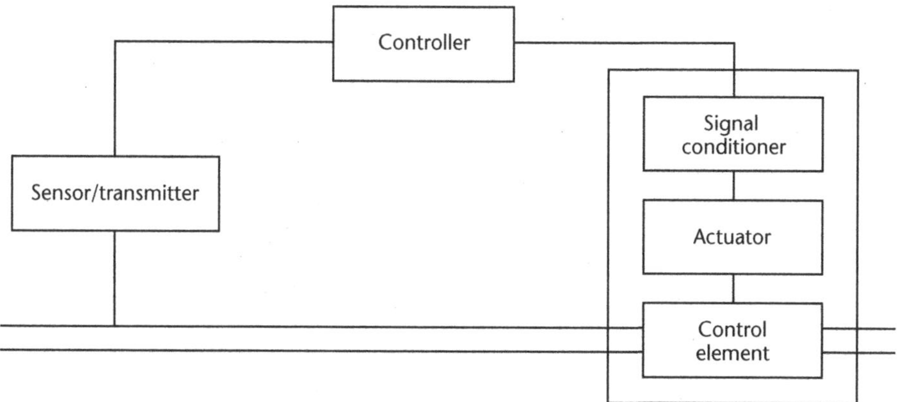

Figure 5-10 Components of control

make the correction. Figure 5-11 illustrates this method of control, for which a chlorinator feed rate is automatically controlled (paced) in response to a signal from a flow transmitter.

Because feedforward control requires the ability to predict the output, this type of control is sometimes called predictive control. Furthermore, because feed-forward control does not measure or verify that the result of the adjustment is correct, it is also referred to as open-loop control. A consequence of open-loop control is that if the measurements, calculations, or adjustments are wrong, the control loop cannot correct itself. In the example in Figure 5-11, there is no check built into the control system that the treated flow of chlorine residual is actually at the desired concentration.

A feedback control loop measures the output of the process, reacts to an error in the process, and then adjusts an input to make the correction. Thus, the information loop goes backward. Figure 5-12 illustrates the control of a chlorinator in response to a residual analyzer. The analyzer continuously adjusts the feed rate of the chlorinator to provide the desired residual in the treated flow. It automatically adjusts for changes in flow rate or chlorine demand.

Because the process reacts only to an error, it is also called reactive control. Furthermore, because feedback control checks the results of the adjustment, it is said to be closed-loop control. Thus, feedback control, unlike feed-forward control, is self-correcting. If the initial adjustment in response to changed conditions does not produce the correct output, the closed-loop system can detect the problem and make another adjustment. This process can be repeated as often as necessary until the output is correct.

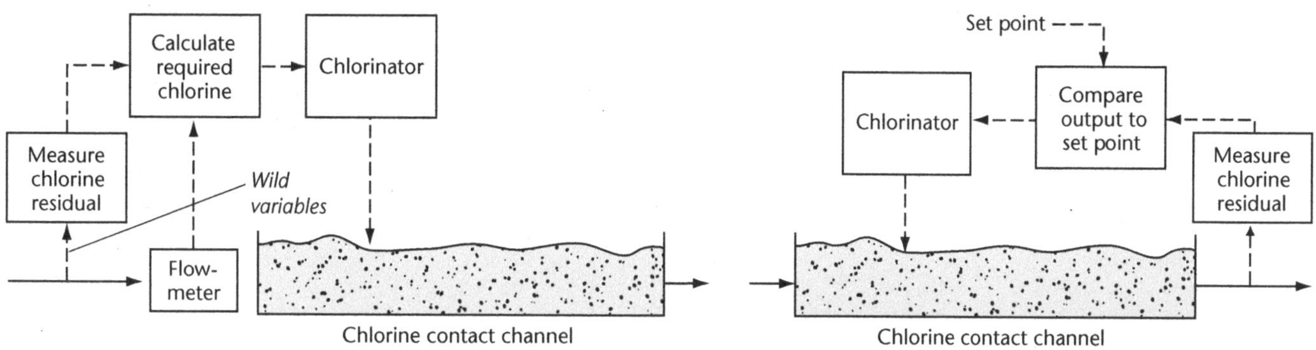

Figure 5-11 Feedforward control of chlorine contact channel

Figure 5-12 Feedback control of chlorine contact channel

Direct Wire and Supervisory Control

Within a plant or an attended pump station, equipment is usually connected to the central control panel through electric wiring. This approach is known as direct-wire control. When a remote station is unattended, the station equipment is controlled from an operator's control location by supervisory control equipment, which transmits control signals over telemetry channels. The unattended site is sometimes called the outlying station, and the operator's station may be called the dispatch station. Many large utilities have a large, central facility to monitor and control the entire water distribution system.

Supervisory Control and Data Acquisition

In a supervisory control and data acquisition (SCADA) system, the control can be remote or automatic. SCADA subsystems consist of the following:

- RTUs
- Communications (telemetry transmission)
- A master station
- Human–machine interface (through graphical format at a central console)

The first two subsystems have already been discussed in this chapter.

Master Station

The main function of a master station includes scanning the RTUs, processing the data, transmitting operator commands, and maintaining a database of historical data (such as valve positions, flow, and pressure). A master station consisting of a single computer is called a centralized station. A distributed system consists of several computer control devices.

Centralized Computer Control

The earliest SCADA systems consisted of a single master station controlling several RTUs. A problem with this system is its dependency on a single computer installation and the communications links to the equipment. For instance, if a remote pumping station is completely operated by a central computer, loss of the computer or the telephone line to the station could mean complete loss of use of the station.

Distributed Computer Control

Smaller, more powerful, and less expensive computers have made it possible to integrate computer control for subsystems of RTUs as well as for individual pieces of equipment. Should the communications link with the central computer be lost, each remote pumping station can be operated by its own computer. Smart equipment (computer controlled) can adjust themselves and monitor their own operation. This reduces, or eliminates, the load on the central computer and decreases the system's dependency on that computer for operation.

Human–Machine Interface

This interface is the connection between the operator and the computer or SCADA system, usually through a central console, keyboard, and mouse. SCADA systems allow personnel with no programming knowledge to set up the display. Operators can design graphics and tabular displays that precisely

direct-wire control
A system for controlling equipment at a site by running wires from the equipment to the onsite control panel.

supervisory control and data acquisition (SCADA)
A methodology involving equipment that both acquires data on an operation and provides limited to total control of equipment in response to the data.

meet their needs. These displays may be interactive. For example, the symbol for a pump may change color or shape depending on status or a reservoir icon may "fill" as the reservoir level increases. Figure 5-13 shows a typical computer control center.

The Future of Supervisory Control

Few water systems exactly fit into any one category of control system. Some very small systems are still primarily operated manually, but there are more that are almost completely automated. Most water systems are somewhere in between. It is clear that the future will bring increasing automation.

State regulatory agencies have mixed emotions about allowing complete automation of water systems. On the one hand, automation may be better than manual control in some ways. It is usually more precise and it eliminates the possibility of human error that is always possible with manual control. On the other hand, although computers and computer programs are becoming increasingly reliable, a simple "bug" in the computer program could seriously damage equipment or cause contamination of the water supply. State regulators will always want to make sure that every automated system is supplied with all possible auxiliary monitoring. The system must be positively designed to shut down or take other appropriate action if anything goes wrong. For more details, see AWWA Standard G100 *Water Treatment Plant Operations and Management.*

One of the newer innovations in computer control is the expert system, which is an interactive computer program that incorporates judgment, experience, rule of thumb, intuition, and other expertise to provide advice about a variety of tasks. In short, the computer is programmed to attempt to copy the decision-making process that would be employed by a human expert.

Such systems are still in the early stages of development and are extremely complex. However, they are seen as eventually having a place in water utility operation by providing sophisticated control over certain operations, resulting in safer and more economical operation.

Figure 5-13 Central computer command center

Operation and Maintenance

Instrumentation represents about 8 percent of the capital investment in mechanical equipment for the average distribution system. Although this may not be a large sum, the proper operation of controls can make a substantial contribution to cost savings by improving overall system performance.

Most new equipment is much more reliable and requires less maintenance than instrumentation of just a few years ago. There are, however, numerous routine maintenance tasks that an operator can perform to keep the instrumentation functioning properly. An instrument's useful life can be increased significantly if it is routinely checked to be sure it is not exposed to moisture, chemical gas or dust, excessive heat, vibration, or other damaging environmental factors.

Maintenance of Sensors and Transmitters

Sensors often require routine maintenance that can be performed by an operator. Special skills may be required to work on transmitters.

Pressure Sensors

Every sensor that in any way responds to liquid pressure will perform poorly if air enters the sensor. Air or gas may enter into the sensing element when released from the fluid line. A vent should be provided on the mechanism and used on a regular basis to allow any accumulated air to escape. A maintenance schedule can eventually be developed as operators gain experience working with the particular sensors. Wherever possible, the sensor should be installed so that trapped gases flow into the upper part of the monitored line, rather than into the sensor.

In cold climates, freezing is a possibility. Most pressure sensors either have provision for a heater and thermostat or can be equipped with special protective cabinets. Heater tape can also be used for protecting sensing lines.

The small pipe connections used to connect pressure sensors, switches, and gauges to the pipeline can be a continual source of trouble where dissimilar metals (e.g., cast iron and brass or bronze) are used together. A valve cock should be installed between the pipeline tap and the device to allow easy disconnection and service. At least twice a year, the sensor should be removed and the valve cock blown and rodded out if necessary. Corrosion of nipples should be checked, and if any weakness is apparent, the failed part should be replaced.

Flowmeters

Several types of flowmeters are used in distribution systems, and each has its own particular maintenance requirements. A venturi tube, orifice, or flow nozzle type of meter has small ports that connect the process fluid to the transmitter mechanisms. These parts should be blown out periodically.

Propeller meters are all-mechanical devices. Over time, wear produces lower readings. The manufacturer's recommendations for lubrication should be carried out regularly, and only the recommended lubricant should be used. A magnetic meter is an electrical unit, so regular checks should be made for corrosion or insulation breakdown around conduit connections or grounding straps.

Transmitters

The operator with no special training in transmitter repair can generally do little more in the way of maintenance than to ensure a favorable environment for the equipment. In hot climates, it is important to protect electronic transmitters from exposure to high temperatures. Most electronic transmitters are rated for operation up to 130°F (54°C). Beyond this temperature, many electronic components may break down. Direct exposure to sunlight may contribute to serious heat buildup within units, and it is often necessary to provide a shield or fan if components are confined in a cabinet. Most transmitters are splash-proof and will function with occasional exposure to wetting. However, unless otherwise specified, most units will not withstand being submerged in water. Therefore, metering pits should have pumps and level alarms to warn of flooding.

Maintenance of Receivers and Indicators

Receiving and indicating devices usually require special skills for service and maintenance. However, the operator should attempt to maintain a favorable environment for the equipment. A dirty or damp atmosphere may damage receivers. Vibration, chemical dust (such as fluoride), and high chlorine content in the atmosphere should all be avoided. The high temperatures that affect transmitters also cause problems for receivers, and inking systems on recording indicators may perform poorly in the cold.

Troubleshooting Guidelines

Some general guidelines for troubleshooting instrumentation systems are as follows:

- Never enter the facility with tools of any type. Most of us have a tendency to start taking things apart without a thorough diagnosis.
- Make a complete diagnosis and attempt to confirm it by example.
- Consider ways to deal with the safety risks.
- Consider what will happen to the operating system while maintenance is being performed. In particular, where control valves are involved, consider the possibility of water hammer.
- Always inform the proper authorities of planned actions beforehand.

Maintenance Records

The development of a maintenance file on instrumentation has many long-term benefits for any distribution system, including the following:

- Accurate accounting of the cost
- Indication of areas where there may be a need for a change in the type of equipment
- Projection of the life expectancy of each type of equipment
- Knowledge required to maintain parts inventories at minimum cost (based on operating experience)
- Proper parts inventories that can reduce emergency downtime substantially (another form of cost savings)
- Guidance for preventive maintenance steps

Many facilities have computerized maintenance management systems or asset management systems that compile data for each type of equipment to track and alert staff of preventive measures that need to be performed in order to maximize equipment life.

Maintenance records should be kept for each instrument. A common problem for maintenance program continuity is the loss of instruction manuals and calibration sheets. A file should be provided for these items, and care should be

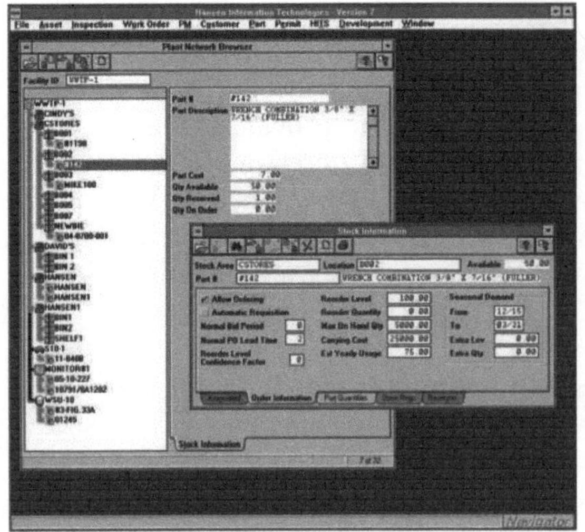

Figure 5-14 Computerized maintenance management system
Courtesy of Hansen Information Technologies.

taken to avoid losing them, especially when responsible personnel are transferred or terminated. Many instrument instruction manuals can be easily accessed online and saved electronically.

Records should also be kept on services related to instruments not owned by the utility, such as telephone lines and power supplies to pump stations. Such information should be carefully documented so that there is concrete evidence in case it becomes necessary to register a complaint. System diagrams showing the routings for communication links are important. For telemetry systems, it is useful to have a map showing channels or frequencies in each tone spectrum and the facilities to which they are connected.

Study Questions

1. What are the basic components of every telemetry system?
   a. Transmitter, signal, receiver, and indicator
   b. Transmitter, receiver, and indicator
   c. Transmitter, transmission channel, and receiver
   d. Sensor, transmitter, transmission channel, and receiver

2. Most telemetry equipment transmit information by
   a. digital signals.
   b. analog signals.
   c. audio signals.
   d. electrical pulses.

3. Control systems consist of which of the following distinct components?
   a. Signal conditioners and control elements
   b. Signal conditioners, actuators, and control elements
   c. Signal conditioners, actuators, control elements, and indicators
   d. Sensors, signal conditioners, actuators, control elements, and indicators

4. Which type of temperature sensor uses a semiconductive material?
   a. Thermistor
   b. Thermoresistor
   c. Thermocouple
   d. Thermoconductor

5. SCADA systems consist of which of the following distinct components?
   a. Remote terminal units (RTUs), communications, and human–machine interface (HMI)
   b. Sensing instruments, RTUs, communications, and HMI
   c. Sensing instruments, RTUs, communications, master station, and HMI
   d. RTUs, communications, master station, and HMI

6. What is the electronic standard range?
   a. 4–20 mA DC
   b. 4–20 mA AC
   c. 0–100%
   d. 0–1 binary

7. In what type of system control is control achieved entirely by machines or computers, without human intervention, under normal conditions?

8. In what type of system control is control achieved by a person operating the switches and levers to control equipment from the physical location of the equipment?

9. What term refers to a type of signal in which the electromotive force varies as the parameter being measured varies?

10. What device consists of a coil of wire that moves when a current is passed through it, activating mechanical levers or switches?

11. What is the term for a technique of monitoring several instruments over a single communications channel with a receiver that periodically asks each instrument to send current status?

Motor Protection Equipment

Some of the automatic relays used to control and protect motors are described in the following paragraphs.

Thermal-overload relays are provided on starters to prevent the pump motor from burning out if abnormal operating conditions increase the load beyond the pump's design capacity. When current is excessive, the relays (sometimes called heaters) placed on each phase of the power supply open the control circuit and stop the motor. These relays are normally set to stop the motor when current exceeds the design load by 25 percent.

Fuses or circuit breakers are placed in the main power wiring to each motor to protect against short circuits. They are normally located in the safety switch just ahead of the starter.

Overcurrent relays are often referred to as overload relays. Their purpose is to sense current surges in the power supply and to disconnect the motor if a surge occurs. They can be fitted with time-delay or preset thermal-overload mechanisms. The overload relays may reset automatically or they may require manual resetting after they are tripped.

Lightning surge arresters are installed at many pump stations to prevent serious damage by high-voltage surges that can be caused when lightning strikes the power line. A surge arrester is usually installed in the service entrance between the power lines and the natural ground. The arrester acts as an insulator to normal voltage; it automatically becomes a low-resistance conductor to ground when the line voltage exceeds a predetermined amount.

Voltage relays are frequently used to detect a loss of power and to initiate a switchover to an alternate power source. Under-voltage relays are also used to shut down motors if the voltage drops too low. Voltage relays generally have a timing mechanism that allows minor functions (ones that won't damage the motor) to continue before power is actually disconnected from the motor.

Frequency relays respond to changes in the frequency (in cycles per second) of an alternating current (AC) power supply. They are most often used where local power generation is involved. Frequency relays are also used on synchronous motor starters to sense when the motor has reached synchronizing speed.

Phase-reversal relays are installed to detect whether any two of the three lines of a three-phase power system are interchanged. If the phase sequence should be reversed, all motors will run backward. A phase-reversal relay senses this change and opens the control circuit to disconnect the motors. A phase reversal can be particularly serious for deep-well pump applications. The pump shafting can become unscrewed, allowing the pump to fail.

Loss-of-phase relays are installed in most wiring systems to detect the loss of one of the three phases. Loss of one phase is not unusual, and if the pump power is not cut off, the motor may burn out within a few minutes while operating on only two phases.

Differential relays are frequently used on large equipment or switchgear. These units check whether all of the current entering a system comes back out of the system. If it does not, the relay closes a contact that shuts down the equipment. An equipment shutdown by a differential relay indicates major trouble, so an electrician should always be called.

Reverse-current relays sense a change in the normal direction of current or power flow and activate an alarm for the operator. They can also open circuits to isolate the faulty portion of the system.

Time-delay relays are used when some condition needs to last for a specified length of time before some other action is begun. For example, if a pump motor must be given sufficient time to reach full speed before the discharge valve opens, a time-delay relay can be energized when the motor is started. The relay can be set to close its contacts several seconds later to activate the valve-opening control circuits.

Bearing-temperature sensors can be placed on any pump or motor to monitor bearing temperature. The sensors incorporate a contact that opens or closes at a predetermined temperature. This action can be used to sound an alarm and initiate an emergency electrical control shutdown.

Speed sensors are used to trigger an alarm or emergency shutdown if a motor should for any reason exceed its normal speed. The sensors are also used on variable-speed motors.

Improving the Efficiency of Electrically Driven Pumps

The power consumed by electrically driven pumps is one of the most costly items in many water utility budgets. Power costs can be reduced in three major ways:

1. Total power use can be reduced by an increase in system efficiency.
2. Peak power use can be reduced if the pumping load is spread more evenly throughout the day.
3. Power-factor charges can be reduced.

Although most major reductions in power costs require replacing equipment, some pumping schedule changes can reduce costs with little or no investment.

Reducing Total Electric Power Usage

In addition to the obvious steps of maintaining motors and pumps in peak condition, running the pumps at the peak of their efficiency curve will reduce power usage. This can readily be accomplished in systems that have multiple pumps.

Where system pressure is higher than necessary for fire and other demands, pressure reduction can reduce the load on pumping facilities. Utilities can often reduce pressure by using high-service pumps in areas where fire protection is critical or by having two-speed motors that are pressure-activated. A reduction in system pressure from 85 to 75 psi (590–520 kPa) could result in energy savings of 5–25 percent, depending on the system flow and pump arrangement.

Reducing Peak-Demand Charges

Unfortunately, water systems and electrical systems experience peak demands during the same periods—early mornings, late afternoons, and the hottest afternoons and evenings of the summer season. The extra capacity installed by electric utilities to provide for peak-demand periods is paid for by demand charges, which high-demand customers pay in addition to the usual rate per kilowatt-hour of electricity.

The demand charge for a billing period is generally calculated based on the average or peak kilowatt demand during the billing period. Demand charges can be minimized if electricity is used at a minimal, constant rate—i.e., if the peak demand is kept as low as possible. Demand charges will also be lower if the water utility's peak demand occurs during a period that is off-peak for the electric utility.

Methods that can be used to reduce demand charges include the following:

- Using gravity-feed storage
- Using engine-driven pumps
- Reducing peak water usage
- Changing the time of demand

Using Gravity-Feed Storage

Gravity-feed storage is an effective way to smooth demand, as well as minimize friction head and reduce overall energy use. For a water utility, the energy conservation that gravity storage provides in meeting peak demand can be substantial. By providing sufficient storage to minimize variations in the pumping rate, the water utility can minimize peak electrical demand and the corresponding demand charges.

Using Engine-Driven Pumps

Using engine-driven pumping units instead of electric-motor–driven units during periods of peak demand may substantially reduce electricity usage and demand charges. This approach may be more energy efficient than installing electrical generators to power pump motors during periods of peak demand. In some areas, natural gas could be used economically to power engine-driven pumps, because summer is the off season for natural gas. Engine-driven units should be operated periodically under load anyway, so a good time to run them is during peak flow conditions.

Reducing Peak Water Usage

Utilities can reduce peak residential water usage to some degree by lowering service pressure, using flow restrictors, or instituting pricing policies that discourage the use of lawn sprinklers. However, the reduction that can actually be accomplished is usually quite small.

Changing the Time of Demand

The methods discussed for reducing peak electrical demand also apply to changing the time of demand to the time of day when off-peak electrical rates are available. Additional gravity-feed storage may be used to permit increased off-peak pumping.

Power-Factor Improvement

The power factor of a motor indicates how effective the motor is at using the power available to it. A value of 1.0 is the highest possible. A power factor of less than 1.0 does not indicate that the motor is inefficient. Instead, it means that the motor requires larger power lines and transformers than are needed for a motor with a power factor of 1.0. For example, a 50-kW motor with a power factor of 1.0 draws 50 kW·h of energy during an hour. The electric utility needs to supply 50 kW of power to the motor. A 50-kW motor with a power factor of 0.8 must be supplied with 50/0.8 = 62.5 kW of power to operate properly (even though the motor will actually use the same amount of energy [50 kW·h] in an hour).

The electric utility imposes a charge for the lower power factor, because its facilities must be able to produce and supply the full 62.5 kW. Power factors vary with the type of motor and the load imposed on the motor. They can also be affected by other components in the motor circuit. Methods of improving the power factor include the following:

- Changing the motor type
- Changing the motor loading
- Using capacitors

Changing the Motor Type

Premium-efficiency motors have higher power factors than standard motors. When purchasing a new motor, a utility should carefully consider the cost savings that might be achieved over time with a motor that is initially more expensive.

Changing the Motor Loading

Although induction-motor efficiencies remain relatively high between 50 and 100 percent of rated horsepower, the power factor decreases substantially and continuously as the rated load falls below 100 percent. For example, at 100 percent of rated load, the power factor of an induction motor may be 80 percent. At 50 percent of rated load, the power factor may be 65 percent, and at 25 percent of rated load, it may drop to 50 percent. Therefore, although reducing the load on motors may decrease power usage somewhat, this advantage is offset to some extent by the reduction in power factor.

Using Capacitors

A utility may improve the power factor of an induction-motor circuit by installing a capacitor in parallel with the motor. The power-factor improvement resulting from a certain capacitance depends on the motor characteristics. Power-factor improvement beyond a certain point may cause problems to an electrical system, so the motor manufacturer should be consulted to determine the capacitance that may be used with a particular motor.

Maintenance of Electric Motors

A preventive maintenance schedule should be developed and followed for all motors in the system. Motors should be inspected at regular intervals, usually once a month. Under severe conditions, more frequent inspections are advisable. Log cards or electronic records should be maintained for each motor. Records should list details of all inspections carried out, maintenance required, and general conditions of operation. These records allow the operator to check the motor's condition at a glance. Any maintenance required should immediately be apparent.

Regular annual or semiannual in-depth inspections are better than frequent casual checks. The following items should be checked during an inspection:

- Housekeeping
- Alignment and balance
- Lubrication
- Brushes
- Slip rings
- Insulation
- Connections, switches, and circuitry
- Phase imbalance

Housekeeping

Routine housekeeping includes keeping the motor free from dirt or moisture, keeping the operating space free from articles that may obstruct air circulation, checking for oil or grease leakage, and routine cleaning. If the correct motor enclosure has been chosen to suit the conditions under which the motor operates, cleaning should be necessary only at infrequent intervals. Dust, dirt, oil, or grease in the motor will choke ventilation ducts and deteriorate insulation. Dirt or moisture will also shorten bearing life.

Alignment and Balance

The alignment and balance on new motors or motors that have been removed and replaced must be carefully checked. They should be rechecked semiannually. Operators should perform the check according to the manufacturer's instructions, using thickness gauges, a straightedge, and/or dial indicators.

The most likely causes of vibration in existing installations are imbalanced rotating elements, bad bearings, and misalignment resulting from shifts in the underlying foundation. Therefore, vibration should always be viewed as an indication that other problems may be present. Both the magnitude and the frequency of vibration should be measured and compared with the measurements made when the unit was first installed.

Lubrication

The oil level in sleeve bearings should be checked and replenished as needed (every 6 months or so). The type and grade of oil specified by the manufacturer must be used. The oil ring, if fitted, should be checked to ensure that it operates freely. Oil wells should be filled until the level is approximately ⅛ in. (3 mm) below the top of the overflow. Overfilling should be avoided.

Grease in ball or roller bearings should be checked and replenished when necessary (usually about every 3 months) with grease recommended by the manufacturer. Bearings should be prepared for grease according to manufacturer's instructions.

New motors fitted with ball or ball-and-roller bearings should be supplied with the bearing housing correctly filled with grease. The bearings should be flushed and regreased every 12–24 months or so. The type of grease recommended by the motor manufacturer should be used. If the motor operates in an environment with a high ambient temperature, a grease with a high melting point should be selected. Where the ambient temperature varies considerably throughout the day, or where there are climatic changes, the manufacturer should be asked for specific advice on the type of grease to use.

It is usually convenient to regrease the bearings after removing the old grease and cleaning the bearings. The operator can perform these tasks by taking off the outside bearing covers when the motor is dismantled for periodic cleaning. The bearing housing should not be overfilled, and the operator should take care that no grit, moisture, or other foreign matter enters the dismantled bearings or the housing. A grease gun can be used for motors fitted with grease plugs.

To prevent deterioration of the insulation, oil or grease should not be allowed to come into contact with motor windings. In all motors fitted with ball bearings, roller bearings, or both, the bearing housings are fitted with glands to keep the grease in and the dirt out. However, over-lubrication of the bearings will result in grease leakage. An overgreased bearing tends to run hot, causing the grease to melt and creep through the gland and along the shaft to the windings. Over-lubricated sleeve bearings can also contaminate motor windings.

Brushes

Brushes should be checked quarterly for wear. They should be replaced before they wear beyond the manufacturer's recommended specification. For the three-phase motors commonly used by water utilities, brushes generally should not be allowed to wear below 0.5 in. (12 mm).

New brushes should be bedded in to fit the curvature of the slip rings. Operators can accomplish this by placing a strip of sandpaper between the slip rings and brushes, with the rough surface toward the brushes. Emery cloth should not be used because the material is electrically conductive and may contaminate components. Normal brush-spring pressure will hold the brushes against the sandpaper. The sandpaper should be moved under the surface of the brushes until the full width of each brush contacts the surface of the slip ring. Where the motor rotates in one direction only, it is best to move the sandpaper in this direction only and then release the brush-spring pressure when the sandpaper is moved back for another stroke.

All carbon dust should be removed after the bedding-in process is complete, including dust that may have worked between the brushes and brush holders. Brush-spring tension should be set as specified by the manufacturer. A small spring balance is useful for checking spring tension. The brush must slide freely in the brush holder, and the brush spring should bear squarely on the top of the brush.

Slip Rings

When properly maintained, slip rings acquire a dark glossy surface with complete freedom from sparking between slip rings and brushes. Sparking will quickly destroy a slip ring's surface, and rapid brush and slip-ring wear will result. Sparking may result from excessive load (starting or running), vibration (e.g., from the driven machine), worn brushes, sticking brushes, worn slip rings, or an incorrect grade of brushes. Periodic inspection and checking will ensure that trouble is reduced to a minimum. If inspection reveals slip rings with a rough surface, the rotor must be inspected for wear and horizontal movement, then repaired as necessary.

Insulation

Motor-winding insulation should be checked periodically to make sure it remains uncontaminated with oil, grease, or moisture. In general, insulation should exhibit at least 1 megohm of resistance.

brushes
Graphite connectors that rub against the spinning commutator in an electric motor or generator, connecting the rotor windings to the external circuit.

Connections, Switches, and Circuitry

All electrical circuits *must* be de-energized and "locked out" before any inspection or maintenance is performed.

Wiring should be checked semiannually or annually. Wires should be examined, and all connections should be checked to ensure that they have not worked loose. Continuous vibration and frequent starting will gradually loosen screws and nuts. Vibration can also break a wire at the point where it is secured to a terminal.

Circuit-breaker and contactor contacts should be examined and cleaned regularly. Slightly pitted and roughened contact surfaces that fit together when closed should not be filed. The burned-in surfaces provide a far better contact than surfaces that have been filed down, no matter how carefully. Any beads of metal that prevent the contacts from completely closing should be removed. Contactor retracting springs should be examined and tested to ensure that they retain sufficient tension to provide satisfactory service.

Moving parts should be operated by hand to check the mating of main, auxiliary, and interlocking contacts. Faces of the contactor armature and holding electromagnet should be clean and should fit together snugly. Anything on the faces that prevents an overall contact when the contactor is closed is likely to lead to chatter and excessive vibration of the mechanism, which increases wear and the possibility of a failure.

Moving parts must be kept clean to avoid any tendency to stick. Check the settings to make sure the adjustments have not been moved by vibration or altered by unauthorized persons. Oil dashpots used with magnetic current relays may need to be replenished with oil of the correct grade in order to maintain the required triggering time delay.

Phase Imbalance

Phase imbalance is caused by defective circuitry in the electrical service. It is the principal cause of motor failures in three-phase pumps. Phase imbalance produces a large current imbalance between each phase or leg of a three-phase service. This imbalance, in turn, will produce a reduction in motor starting torque, excessive and uneven heating, and vibration. The heat and vibration will eventually result in the failure of the motor windings and bearings.

Some phase imbalance will occur in any system but it generally should not exceed 5 percent. Each phase value should be checked monthly with a volt-ammeter for balance, and the values should be logged. If excessive imbalance is found, an electrician or the power utility should be called to isolate and repair the problem.

When a fuse blows on one leg of a three-phase circuit, a phase-imbalance situation known as single-phasing occurs. The fuse may have blown because of an insulation failure in the motor or because of problems with the power line. Unless the motor is manually or automatically turned off, it will continue to try to run on the two remaining phases. However, the windings that are still under power will be forced to carry a greater load than they were designed for, which will cause them to overheat and eventually fail.

Motor and Engine Safety

Operators must follow special safety precautions when dealing with motors and engines. In addition to all the other safety concerns associated with water distribution (as discussed in other chapters), operators must be cautious around electrical devices and be aware of fire safety guidelines.

Electrical Devices

There is no safety tool that will protect absolutely against electrical shock. Operators should use plastic hard hats, rubber gloves, rubber floor mats, and insulated tools when working around electrical equipment. However, these insulating devices do not guarantee protection, and the operator using them should not be lulled into a false sense of security.

Electrical shocks from sensors are possible in many facilities, such as pumping stations, because many instruments do not have a power switch disconnect. It is important to tag such an instrument with the number of its circuit breaker so that the breaker can quickly be identified. After the circuit breaker has been shut off, an operator should tag or lock the breaker so other employees will not reenergize the circuit while repairs are being performed. Even after a circuit is disconnected, it is good practice to check the circuit with a voltmeter to be certain that all electrical power has been removed.

It is very easy to damage an electrical or electronic instrument by inadvertently shorting a circuit while making adjustments. Insulated screwdrivers should be used for electronic adjustments to reduce the chance for damage.

Blown fuses are usually an indication of something more than a temporary or transient condition. The cause of the overload should be identified. Never replace a blown fuse with a fuse of higher amperage than the circuit's designed rating.

Broken wires should be replaced instead of repaired. Secure terminal connectors should be used on all wires at all points of connection. Charred insulation on a wire is a warning of a serious problem.

Extreme care should be taken in working around transformer installations. Maintenance activities should be performed only by authorized employees of the power company. Power to a transformer cannot be locked out except by the power company, so operators should always assume that all transformers are energized and observe full safety precautions.

Electric switchboards should be located and constructed in a manner that will reduce the fire hazard to a minimum. They should be located where they will not be exposed to moisture or corrosive gases, and their location should allow a clear working space on all sides.

Adequate illumination should be provided for the front (and back, if necessary) of all switchboards that have parts or equipment requiring operation, adjustment, replacement, or repair. All electrical equipment, including switchboard frames, should be well grounded. Insulating mats should be placed on the floor at all switchboards.

Open switchboards should be accessible only to qualified and authorized personnel and should be properly guarded or screened. Permanent and conspicuous warning signs should be installed for panels carrying more than 600 V. Areas screened off because of high voltage should be provided with locks to prevent unauthorized persons from entering. However, the lock must be operable by anyone from the inside so that there is no chance of someone being locked inside the enclosure.

Switches should be locked open and properly tagged when personnel are working on equipment. Fully enclosed, shockproof panels should be used when possible. Such equipment should be provided with interlocks so that it cannot be opened while the power is on.

Fire Safety

Suitable fire extinguishers should be kept near at hand and ready for use. The locations and operation of fire extinguishers should be familiar to all employees. Water or soda-acid extinguishers should never be used on electrical fires or in the vicinity of live conductors. Carbon dioxide or dry-powder extinguishers are recommended for these situations.

Detailed fire safety requirements vary considerably from one installation to another and from one locality to another. Therefore, it is recommended that the advice (and where necessary, the approval) of one or more of the following organizations be secured: local fire department, state fire marshal's office, fire insurance carrier, or local and state building and fire prevention bureaus. Generally, the fire insurance carrier and one of the fire protection agencies will provide all of the necessary advice.

Study Questions

1. The current drawn by a motor the instant it is connected to the power source, called the locked-motor current, is usually __________ times the normal full-load current of the motor.
   a. 1.5–2
   b. 2–3
   c. 3–5
   d. 5–10

2. Which system would provide for a "soft start" to a motor?
   a. Wound-rotor induction motor and a controller
   b. Synchronous motor and a variable-frequency controller
   c. A variable-frequency drive and a squirrel-cage induction motor
   d. Squirrel-cage induction drive and a variable-frequency drive

3. Which type of relay is frequently used to detect power loss and initiate a switchover to another power source?
   a. Loss-of-phase relay
   b. Voltage relay
   c. Frequency relay
   d. Circuit breaker relay

4. Which type of relay is placed on each phase of a power supply to open a control circuit and stop a motor, if the current becomes excessive?
   a. Thermal-overload relay
   b. Frequency relay
   c. Voltage relay
   d. Differential relay

5. Which type of relay is often used where local power generation is involved and on synchronous motor starters to sense when the motor has reached synchronizing speed?
   a. Speed sensor
   b. Differential relay
   c. Frequency relay
   d. Voltage relay

6. Which type of relay is frequently used on large equipment to check whether all the current entering a system comes back out of the system?
   a. Voltage relay
   b. Overcurrent relay
   c. Reverse-current relay
   d. Differential relay

7. Which of the following is *not* an approach to improving a motor's power factor?
   a. Using capacitors
   b. Changing the motor type
   c. Changing the motor loading
   d. Decreasing the size of power lines and transformers

8. Some phase imbalance occurs in any three-phase circuit, but it should not generally exceed what percentage?

9. List three approaches to lowering the cost of power use.

10. What is the highest power factor possible for a motor?

11. Where should fire extinguishers be kept?

12. What is the principal cause of motor failure in three-phase pumps?

Chapter 7
Meters

Customer Meter Installation

Customer meters can be installed either in outdoor meter pits or in the building being served. Installation of meters in basements is quite common. When there is no basement, some utilities allow installation in a crawl space or a utility closet. **Meter pits** (or **meter boxes**) are usually located in the parkway, between the curb and sidewalk.

Large meters are often installed in precast concrete or concrete-block vaults if there is no appropriate location in a basement. Large-meter installations are expensive and require considerable planning.

General Considerations

Indoor installations are more common in northern states, where harsh winter weather can cause frost damage to the meter and where houses are more likely to have basements. Outdoor installations are more common in warmer, more temperate climates. Whether the meter is installed indoors or out, there are several general requirements for acceptable installation, including the following:

- The meter should not be subject to flooding with nonpotable water.
- The installation should provide an upstream and a downstream shutoff valve to isolate the meter for repairs.
- The installation should position the meter in a horizontal plane for optimal performance.
- The meter should be reasonably accessible for service and inspection.
- The location should provide for easy reading either directly or via a remote reading device.
- The meter should be reasonably well protected against frost and mechanical damage.
- The meter installation should not be an obstacle or hazard to customer or public safety.
- The meter should have seals attached to the register to prevent tampering.
- There should be sufficient support for large meters to avoid placing stress on the pipe.
- A large installation should have a bypass or multiple meters so that water service does not have to be discontinued during meter replacement or repair.

meter box
A pit-like enclosure that protects water meters installed outside of buildings and allows access for reading the meter.

Manifold Installations

Some water systems commonly install two or three meters in a manifold, or battery, for customers that require a high flow rate. As illustrated in Figure 7-1, a 4-in. (100-mm) service line can have three 2-in. (50-mm) displacement-type meters installed. Advantages of a battery installation include the following:

- Meters can be removed one at a time for servicing without disrupting customer service. This differs from a single large-meter installation, for which the customer is usually given free water through a bypass while the meter is being repaired.
- Meters can be added and the system easily expanded if required.
- There is no need to buy or stock parts for several different-size meters. All meters and valves are the same size.
- The battery can be mounted with the meters stacked along a basement sidewall to conserve floor space.

In a manifold unit with two or three meters (Figures 7-2 and 7-3), all but one of the meters must have a lightly loaded backpressure valve on their outlets. This

Figure 7-1 Manifold connection of three meters

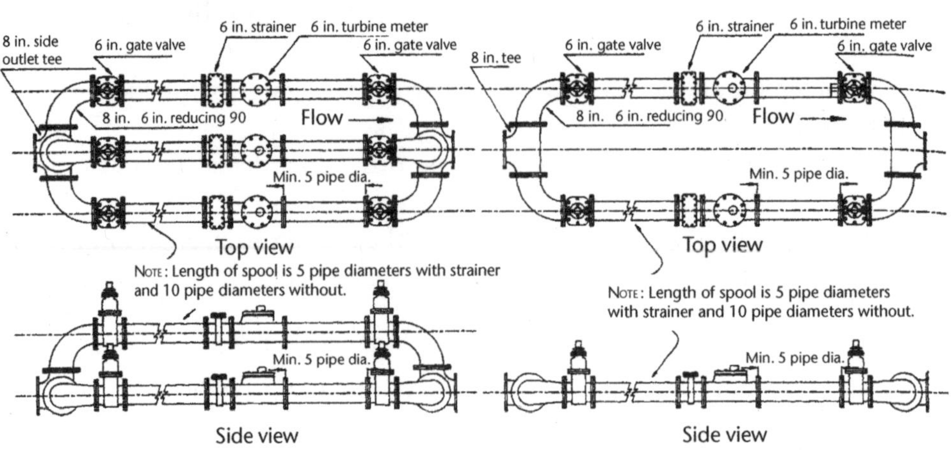

Typical three-meter and two-meter hydraulically balanced manifold.

Figure 7-2 Manifold of large meters

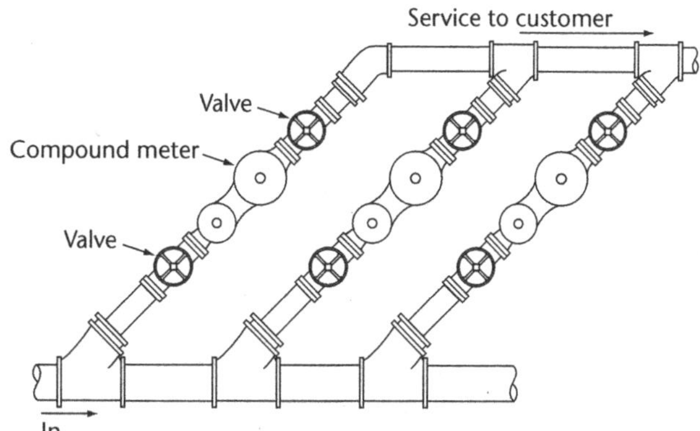

Figure 7-3 Diagram of manifold of three large meters

way, when the flow is small, only one meter will operate. As flow increases, the backpressure valves open to permit flow through the other meters.

The price difference between a multiple-meter installation and a compound meter is another consideration. Older-model compound meters were very expensive, but newer models are available that are more competitively priced.

In the past, commercial and industrial services were quite often metered with compound meters. These meters register well over a wide flow range and have a relatively low pressure loss at high flows. Today, however, new types of horizontal turbine meters should be considered for these customers. They are quite sensitive to low flows, may have a lower initial cost, and may require less maintenance.

Meter Connections

The types of connections used on water meters vary with the size of the meter. Meter sizes up to 1 in. (25 mm) usually have screw-type connections (Figure 7-4), whereas larger meters usually have flanged connections (Figure 7-5).

Many water utilities use a special device called a meter yoke or horn to simplify the installation of small meters. Figure 7-6 illustrates several types that are commonly used for both interior and meter-pit installations. The purpose of a yoke is to hold the stub ends of the pipe in proper alignment and to maintain spacing to support the meter. Yokes also cushion the meter against stress and strain in the pipe and provide electrical continuity when metal pipe is used.

Figure 7-4 A ¾-in. (20-mm) meter with screw connections

Courtesy of Neptune Technology Group Inc.

Figure 7-5 A 1½-in. (40-mm) meter with flanged couplings

Courtesy of Neptune Technology Group Inc.

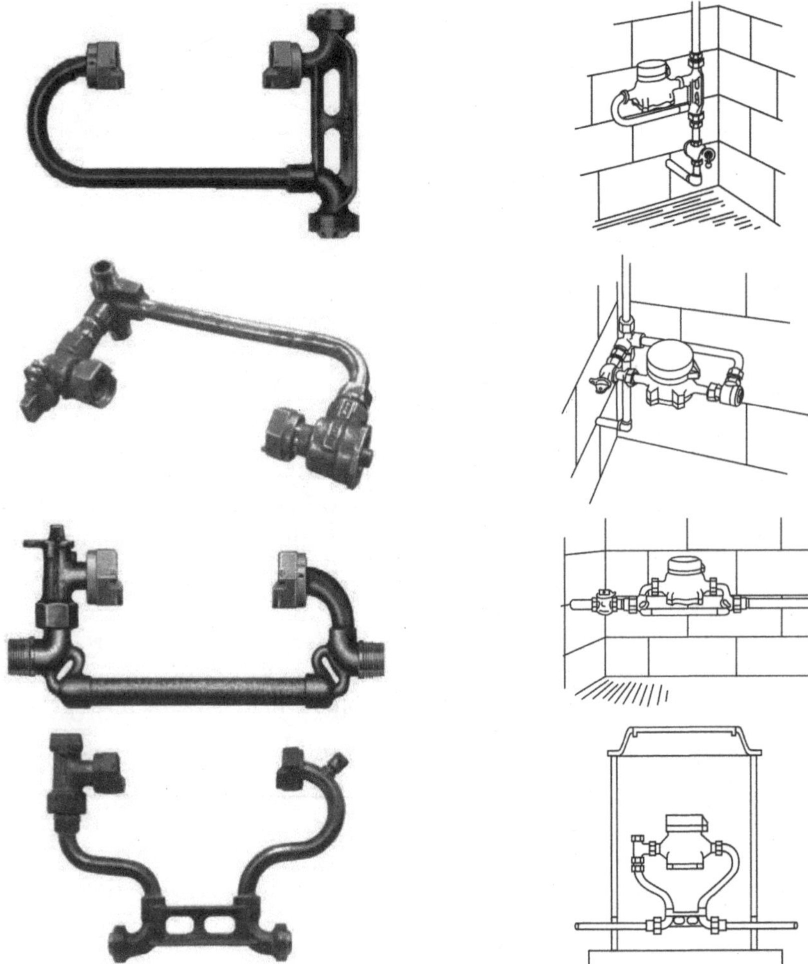

Figure 7-6 Various styles of meter yokes

Courtesy of Mueller Company, Decatur, Illinois.

Indoor Installations

Many water systems have developed a diagram to illustrate how a meter should be installed inside a building under normal circumstances. Ideally, the meter should be located immediately after the point where the service pipe enters through the floor or wall. If meters are to be read directly, the location must be kept relatively clear to allow convenient reading. If meters are to be furnished with remote reading devices, the location should still allow for reasonable access. This will allow periodic direct readings to check on the remote device. It will also allow the meter to be changed when repair is required.

Some installation details should be standardized, including the following:

- Minimum and maximum heights for the meter above the floor
- The types of meter connections to be used
- The required type of valve before and after the meter
- The minimum access space required for reading and servicing

Where the building electrical system uses the water service as a ground, current can flow through the piping and create an electrical hazard to employees who remove the meter or repair the service line. This current also increases the possibility of corrosion of the service line and connections. If the water service pipe is used as a ground and a meter yoke is not used, an electrical ground connection must be installed across the meter. If the water service is plastic pipe, grounding the electrical system to the plumbing system obviously serves no purpose.

Having a diagram of the required meter location and plumbing ensures proper installation for the mutual benefit of the water utility and the customer. A copy of the diagram should be furnished to those applying for water service. A water utility should require compliance with the specifics of the diagram before providing service.

Outdoor Installations

Outdoor meter installation varies depending on the size of the meter involved.

Small-Meter Installation

A meter box or pit is usually required to protect small outdoor meters. Many factors influence the design, materials, installation details, and overall performance of an outdoor setting. Factors include soil conditions, groundwater level, maximum frost penetration, and accessibility for reading and servicing.

The wide variations in ground frost penetration throughout the country make it impossible to detail a universally practical outdoor setting. As with the indoor setting, a diagram for use by water system employees, building inspectors, contractors, and homeowners should be prepared. This diagram should illustrate a standard for meter pit location, construction, and plumbing. The meter pit is usually constructed and furnished by the owner's contractor. It is particularly important that the contractor be furnished with explicit instructions. Because each community has different requirements on how the installation should be made, it is often hard for contractors to remember what each town requires. The following guidelines for the standard diagram should be observed:

- If at all possible, the meter pit should be located on public property or at least relatively close to the property line.
- The location should be relatively safe from possible damage from vehicles and snow-removal equipment.
- The pit or box lid should be tight fitting, tamper resistant, and placed flush with the ground surface so that it does not create a tripping hazard or interfere with lawn mowing.
- Where ground frost is expected, the riser pipes should be 1–2 in. (25–50 mm) away from any portion of the meter box walls to avoid freezing.
- The distance below ground surface at which the meter spuds or couplings are to be located must be specified. In general, small meters are raised to facilitate reading and meter replacement.
- The dimensions of the meter box to be used for each size of meter must be specified.
- The location and type of curb stop or service control valve must be specified.
- A meter yoke is recommended for pit installation to make it easier to change the meter and to prevent distortion of the meter body due to pipe misalignment caused by shifting over time.

Large-Meter Installation

There is no uniform standard for large-meter installations. A prime consideration in their design is that large meters are very heavy and require adequate support so

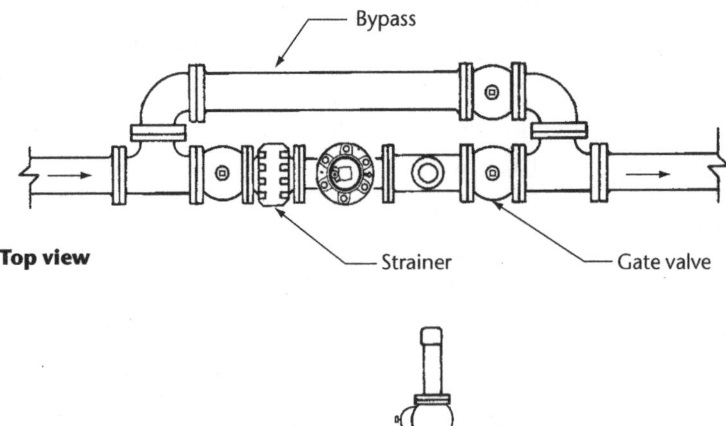

Figure 7-7 Large-meter installation diagram

Courtesy of Neptune Technology Group Inc.

that no stress is put on the service pipe. The cover on the meter pit must be made large enough for worker entry and meter removal.

Adequate work space must be allowed around large meters installed in a vault. At least 20 in. (510 mm) of clearance to the vertical walls and at least 24 in. (610 mm) of head space from the highest point on the meter should be allowed. More space is desirable. Test valves should be installed to permit volumetric tests. Provisions should be made for discharging the test water if meters are to be tested in place. The meter and valves should be supported, and thrust blocking should be provided when necessary. A typical large-meter installation is illustrated in Figure 7-7.

Valuable aid and installation recommendations can be obtained from meter manufacturers. These manufacturers can make recommendations on meter vault designs commonly used in the area, as well as on plumbing materials that will provide the most economical and satisfactory installation.

Meter Reading

Meters used in the United States are generally furnished with **registers** that record the flow of water in gallons or cubic feet. Registers that read in imperial gallons or cubic meters are also available. Water meter registers are usually of two types: circular or straight, as illustrated in Figure 7-8. Circular registers, which are somewhat difficult to read, have gradually been replaced by straight registers on new meters. Straight registers are read like the odometer on a car. The meter reader simply reads the number indicated on the counting wheels, including any fixed zeros to the right of the counting wheel window.

register
That part of the meter that displays the volume of water that has flowed through the meter. Meter registers are generally either of the straight or circular type.

The registers on large meters occasionally have a multiplier on them. If so, the multiplier such as "10×" or "1,000×," will be noted on the meter or the register face. This marking indicates that the reader must multiply the reading by the multiplier in order to obtain the correct reading.

Direct Readout

The most common method of meter reading is the direct readout, which involves an individual going from one meter to another and directly reading the registers.

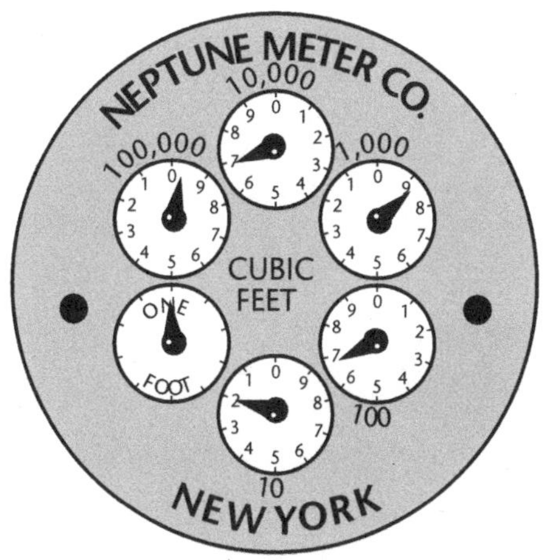

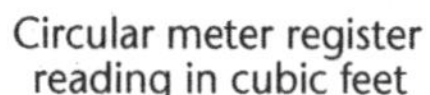
Circular meter register reading in cubic feet

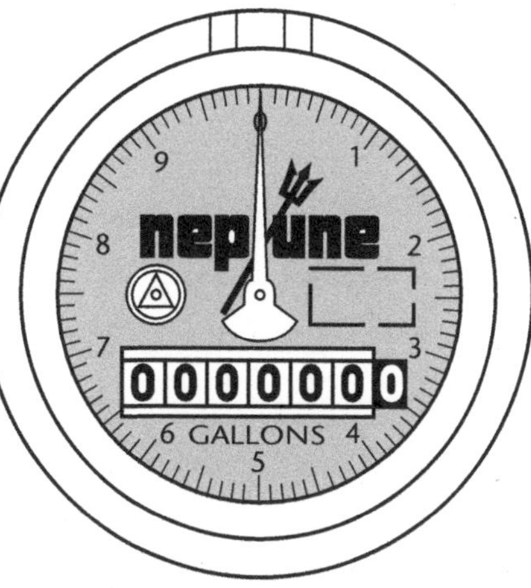

Straight meter register reading in gallons

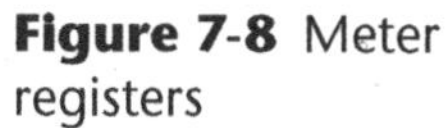
Figure 7-8 Meter registers

Courtesy of Neptune Technology Group Inc.

If the meter is located in a home, the following problems may be encountered:

- Some residents are reluctant to admit the meter reader into their home for a variety of reasons, including fear of being attacked, damage to property, and inconvenience.
- Some meters are in very inconvenient locations, either because the initial location was poorly planned or because the resident has subsequently built something around the meter to hide it.
- Often no one is home during working hours on weekdays.

Water utilities can use alternative methods to obtain readings from homes where the occupants are not home during the week.

- A meter reader can work on Saturday to make special readings from homes missed during the week.
- The utility can make special arrangements by phone to meet residents at a specific time when they will be home.
- A special card can be hung on the doorknob, asking the resident to read the meter and return the card. Some residents who dislike being bothered by the meter reader will ask that a card be left for every reading.

If the meter is located outside, the following problems may arise:

- The meter pit must be located each time a reading is taken. If grass, dirt, snow, or stones have covered it, the reader must spend time finding and exposing the cover.
- With deeper pits, the pit may fill with water at certain times of the year and must be pumped before the meter can be read.
- With large meters or small meters that haven't been raised, the reader may have to crawl into the pit to make the reading.

Remote Reading Devices

Remote reading devices were developed to eliminate most of the problems with direct meter reading inside buildings. The most common type of remote setup transmits the signal electrically (Figure 7-9). The meter register contains a pulse generator that stores energy in a spring-release mechanism. This energy is

Figure 7-9 Remote meter-reading device

Courtesy of Badger Meter, Inc., Milwaukee, Wisconsin.

accumulated very slowly as a specific amount of water—either 10 ft^3 (0.3 m^3) or 100 gal (380 L)—passes the meter. When this point is reached, the energy is released by a mechanism that creates rapid motion between a permanent magnet and a copper coil. This motion generates an electric impulse that is transmitted through a two-conductor electric cable to the remote register mounted at a convenient location on the outside of the building. The remote register contains a counter that advances one digit for each pulse received.

Remote reading devices are quite popular in northern states where most meters are located inside. The use of remote registration has been increasing in areas where pit settings were more common, both in residential and in commercial and industrial installations. In areas where large meters are installed under the street, it is often necessary to send a truck and a crew to direct traffic, manipulate pit entries, and then read the meters. However, conveniently located remote registers make it possible for one person to obtain these meter readings without disrupting traffic or risking personal injury.

When considering the installation of remote registration systems, the water supplier should thoroughly study all related factors. Much of this information, including the following: is available on request from meter manufacturers:

- Comparisons of new meter setting costs with and without remote reading
- Desirability and economics of retrofitting existing installations
- Interchangeability within meter models and brands used by the water supplier, Compatibility of the reading data obtained if both direct and remote systems are intermixed in the same route
- Original equipment purchase price
- Installation costs
- Probable maintenance costs
- Ultimate cost of obtaining and processing meter-reading data

Plug-in Readers

Several manufacturers offer remote meter-reading units that have a plug-in receptacle. With this system, the reading is converted into an encoded electrical message within the meter head. This meter head is connected by a multiconductor cable to a receptacle located on an outside wall of the building. A meter reader can then plug a battery-operated reading device into the remote receptacle to take a reading. The meter reading may be visually displayed and also recorded for later direct entry into a computer.

Remote register or plug-in units are not as advantageous for meters in pits. They are sometimes mounted under the pit cover. Although the cover must be raised to make the reading, this approach is still easier than having a worker crawl into the pit. The alternative is to find a protected spot near the pit where the register can be mounted aboveground, but this is rarely convenient.

Electronic Meter Reading

A "scanning" unit for electronic meter reading (Figure 7-10) works well for either indoor or pit-mounted meters. With this type of unit, an inductive coil is mounted either on the wall of the building or in the cover of the meter pit. It is connected by wires to a special register on the meter. The meter reader carries a unit that includes a probe on an extension arm that simply has to be held in the proximity of the inductive coil to obtain a reading. Power from the interrogator is transmitted through the inductive coil to the register's microprocessor. The meter identification number and reading are transmitted back to the unit carried by the meter reader.

Automatic Meter Reading

Automatic meter reading (AMR) enables a utility to obtain readings without actually going near the meters. Methods include transmitting a meter signal through the telephone system, through electric power distribution networks, via sound transmission through water lines, through cable TV wiring, by radio, or by satellite (Figure 7-11) and cellular networks. If radio transmission is used, different options are available, such as operating the entire system from a central radio tower, collecting readings from radio-equipped trucks that drive down the street, or having the readings obtained by a meter reader who is walking or driving by on the street (Figures 7-12 and 7-13).

In addition to the obvious advantages of avoiding the costs and problems of requiring meter readers to visit each meter, AMR makes it practical to read meters and bill monthly, rather than quarterly as practiced by many water systems. It also makes it easy to perform studies such as determining how much water is being used by various classes of customers on a peak water use day.

automatic meter reading (AMR)
Any of several methods of obtaining readings from customer meters by a remote method. Methods that have been used include transmitting the reading through the telephone system, through the electric power network, through water lines (via sound transmission), through cable TV wiring, and by radio.

Figure 7-10 Proximity meter-reading system
Courtesy of Badger Meter, Inc., Milwaukee, Wisconsin.

Figure 7-11 Remote meter-reading system
Courtesy of Badger Meter, Inc., Milwaukee, Wisconsin.

Figure 7-12 Handheld computer
Courtesy of Neptune Technology Group Inc.

Figure 7-13 Mobile data collector
Courtesy of Neptune Technology Group Inc.

Meter Testing, Maintenance, and Repair

Water meters should be tested before use, when removed from service, when a customer makes a request or complaint, and after any repair or maintenance.

Testing New Meters

New meters are tested by water suppliers to identify damaged meters and limit metering errors. Because of the time involved in testing new meters, some state regulatory commissions have adopted regulations that allow a certain number of random meters in a shipment to be tested to determine the accuracy of the entire shipment. For example, if meters are shipped in boxes of 10, a water supplier might test only 1 random meter in that box. If that meter tests out accurately, all of the meters in the box can be assumed to be accurate. If the meter does not test out accurately, the water supplier would have to test all the other meters in the box or return all of them to the supplier for replacement. The state public utility regulatory commission should be consulted concerning any requirements it may have for testing new meters before use.

Frequency Requirements for Testing Meters

Water meters are subject to wear and deterioration. The rate of wear and deterioration is principally a function of how much the meter is used and the quality of the water. Over time, meter efficiency decreases and meters fail to start operating under low flow. Although some customers may sometimes think otherwise, meters very rarely overregister. Occasionally a customer insists on having an old meter replaced because he or she thinks it is running fast. The usual result is that the new meter will be more accurate and the bill will be higher.

To avoid a loss of revenue for the utility, meters must either be field tested or brought in periodically for testing and repair. To help control meter deterioration and limit registration problems, many state regulatory commissions have adopted requirements for the frequency of meter testing. Because of the variability of water characteristics, the cost of testing, and water costs, a nationwide test frequency cannot be established. Most states with frequency requirements for meter testing base the test intervals on the size of the meter—the larger the meter, the more often it should be tested. AWWA recommends that every utility have a scheduled meter-testing program. The determination of the most appropriate test interval should be based on utility experience, local conditions, and cost/benefit considerations. A statistical sampling approach may be the best choice for some systems while a timed approach may be the most suitable for others. Considerations for determining the testing interval are discussed in AWWA Manual M6, *Water Meters—Selection, Installation, Testing, and Maintenance.*

Whether a water utility tests its own meters or has someone else do it depends on the staff size, time available, and facilities. Although some utilities have the facilities to test all their meters, the majority test only smaller meters. Larger meters may be field tested or sent to the manufacturer for testing. Smaller water systems do not usually have the facilities or the time to test meters, so they send the meters to the manufacturer or to a contractor for testing and repair. The economics of simply disposing of old meters and purchasing new ones must also be considered.

Testing Procedures

There are three basic elements to a meter test:

1. Running a number of different rates of flow over the operating range of the meter to determine overall meter efficiency.
2. Passing known quantities of water through the meter at various test rates to provide a reasonable determination of meter registration.
3. Meeting accuracy limits on different rates for acceptable use.

The rates of flow generally used in testing positive-displacement meters are maximum, intermediate, and minimum. For current and compound meters, four or five flow rates are usually run. State regulations should be consulted to determine if specific rates are required.

Equipment for meter testing does not have to be elaborate. A setup like the one shown in Figure 7-14 is suitable for small meters. It is generally more economical to field-test large meters, especially current and compound types. Field testing is essentially the same as shop testing. The difference is that, instead of using a tank to measure the test water, operators compare the meter to be tested and one that has previously been calibrated. The two meters are connected in series, and the test water is discharged to waste. Because the calibrated meter is not 100 percent accurate at all flows, it is necessary to use a special calibration curve to adjust for different rates of flow in order to compute the accuracy of the meter being tested.

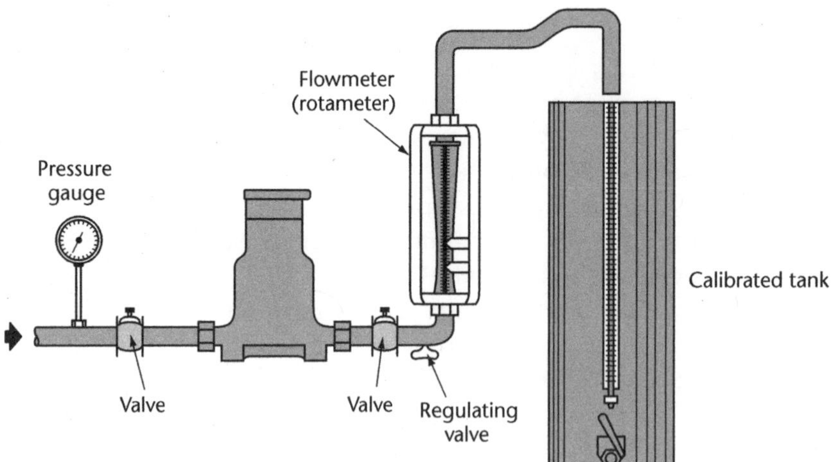

Figure 7-14 Meter-testing equipment

Some water suppliers do not field-test large meters over their entire operating range. Instead they check only the lower 5–10 percent of the rated or operating range. Utilities may take this approach because the large quantities of water discharged in high-flow tests often present a problem. The underlying assumption for this approach is that the test curve will flatten out after reaching a peak registration that is approximately 10 percent of rated capacity. Presumably the meter will then stay within the required limits of registration.

For field testing to be accurate, both meters must be full of water and under positive pressure. The control valve for regulating flow should always be on the discharge side of the calibrated meter. A control valve on the inlet side of the meter configuration or one located between the two meters should not be used.

Maintenance and Repair

Maintenance and repair for positive-displacement meters involves the following general steps:

1. Dismantle the meter.
2. Thoroughly clean the parts.
3. Inspect all parts for wear, pitting, and distortion.
4. Replace or repair parts as necessary.
5. Reassemble the meter.
6. Retest the meter.

Maintenance and repair for other meters generally follow the same procedure, with some variations due to design differences.

Before undertaking a maintenance and repair program, water utilities should evaluate the cost-effectiveness of a meter replacement program versus a meter repair program. It is sometimes more economical simply to replace old parts with new parts or old meters with new meters, rather than to check tolerances, shim disks, or repair registers. Meter manufacturers can furnish detailed repair instructions for each style of meter.

Record Keeping for Meters

Meter records are an essential part of any water distribution system that meters its supply. Meter records should provide information on the installation, repair, and testing of each meter. The records enable field crews to locate meters easily, help repair personnel with meter testing, and aid managers in assessing the value of the system.

One method for maintaining meter records involves having a meter history card for each meter. Basic data recorded at the top of the card include the size, make, type, date of purchase, and location. The remaining portion of the card is divided up to record the location of the meter and, in chronological order, all tests and repair work. Each line of the test and repair section is usually divided into two segments. The upper part is used to record test results when the meter is removed, and the lower part is used to record test results before the meter is reinstalled after maintenance or repair.

In a small- to medium-size shop, test and repair information should be entered on the meter history card by personnel in the meter repair shop. After all information is recorded, meter history cards should be stored in a safe, permanent file in the shop area. They are usually filed in sequence, either by the manufacturer's serial number or by the water system's number.

Many water suppliers use computerized meter maintenance systems in their operation. Information commonly kept on a service or meter history card is entered into the computer system to establish a permanent record. A control number is associated with each service or meter. Any future information concerning work on a customer service line or meter testing and repairs can be entered for the appropriate control number. These systems can be integrated with geographical information systems (GISs) to provide the utility with comprehensive information regarding the meter history, location, service, land ownership, billing, distribution system main and value information, etc.

Study Questions

1. Where ground frost is expected, the riser pipes in a meter pit should be at least how far away from any wall of the meter box to prevent freezing?
   a. 1–2 in. (21–51 mm)
   b. 2–4 in. (51–102 mm)
   c. 4–6 in. (102–152 mm)
   d. 6–8 in. (152–203 mm)

2. What is the main purpose(s) of the meter yoke?
   a. Provide cushion against stress
   b. Provide cushion against stress and strain
   c. Provide electrical conductivity
   d. Provide proper alignment and support for the meter

3. Large meters are usually tested at what percentage of their operating range?
   a. 5–10%
   b. 10–20%
   c. 20–25%
   d. 25–33%

4. When installing a meter, it should be positioned _____ for optimal performance.
   a. in a horizontal plane
   b. diagonal to the ground
   c. away from the house
   d. facing downward toward the ground
5. Registers record the flow of water in
   a. pounds per square inch.
   b. cubic meters.
   c. gallons or cubic feet.
   d. feet.
6. Which of the following is *not* considered an installation detail that should be standardized?
   a. The minimum access space required for reading and servicing
   b. Days of the week on which the meter may be read
   c. Minimum and maximum heights for the meter above the floor
   d. The required type of valve before and after the meter
7. What is the term for the part of the meter that displays the volume of water that has flowed through the meter?
8. Ideally, where should the meter be located?

Chapter 8
Operational Practices

Operations and Maintenance Practices to Maintain Water Quality

Operators can influence water quality through their system operating procedures. These practices need to focus on three issues:

- Minimize bulk water detention time (control water age).
- Maintain positive pressure.
- Control the direction and velocity of the bulk water.

Most water quality problems in the distribution system can be controlled by effectively addressing these concerns.

Hydraulic Detention Time

Water quality deterioration is often proportional to the time the water is resident in the distribution system. The longer the water is in contact with pipe walls and is held in storage facilities, the greater the opportunity for water quality changes. Standard operating procedures (SOPs) should be developed to minimize the detention time in the system. Hydraulic models can be used to help define water age throughout the system and to help evaluate ways to reduce this value. Monitoring for disinfection by-products (DBPs) and disinfectant residual can also provide information leading to the identification of areas where detention time may be excessive.

Storage Facility Operation

Finished water storage facilities have been sized and operated to provide reserves for emergency service and firefighting needs and to satisfy peak demands. These requirements often lead to oversizing the facility for optimum water quality. SOPs should include practices that will promote mixing and reduce water age.

Water quality in storage facilities should be carefully monitored. Routine sampling of the facility outlet is a minimum monitoring step. Operators may also need to periodically examine the water quality at other locations inside the facility, including taking samples at various depths. It is possible for the water to become stagnant or stratified within the facility. Disinfectant residual can be a good parameter to use to evaluate water quality changes in storage facilities.

SOPs should establish a minimum turnover goal based on water quality at the facility. Generally, one complete turnover of the contents of a finished water storage facility is recommended (seasonal adjustment may be necessary) about every 5 days. A fluctuating water level promotes mixing. However, the process used to

detention time
The average length of time a drop of water or a suspended particle remains in a tank or chamber. Mathematically, it is the volume of water in the tank divided by the flow rate through the tank.

fluctuate water level may not always result in the removal of stagnant water. An effective method is to lower the water level in a continuous operation rather than in small increments throughout the day.

Some relatively rare storage facilities need special operational considerations. Facilities with floating covers must take precautions to avoid damaging the fabric. Ice formation is a particular concern. Mechanical recirculation systems are provided to mix water in some facilities. These installations need to be monitored to verify their effectiveness, and the mechanical equipment must be constructed to prevent possible water contamination. Storage facilities are a convenient place to redisinfect the water. This may be done by adding the disinfectant to the inlet, to the outlet, or inside the facility (batch disinfection). Chlorine is most commonly used for this purpose. If the water is chloraminated, special care must be taken so that the desired residual (free chlorine or combined chlorine) can be attained.

Flushing Programs

The velocity of flow in most mains is normally very low (hydraulic detention time can be long) because mains are designed to handle fire flow, which may be several times larger than domestic flow. As a result, corrosion products and other solids tend to settle on the pipe bottom. The problem is especially bad in dead-end mains or in areas of low water consumption.

These deposits can reduce the carrying capacity of the pipe. They can also be a source of color, odor, and taste in the water when the deposits are stirred up by an increase in flow velocity or a reversal of flow in the distribution system. These sediments provide an environment for future biofilm growth and can result in a high disinfectant demand. In some cases, the disinfectant residual can be completely consumed. In areas where other operations cannot improve the hydraulic detention time, flushing can restore the water quality and help avoid the need for reactive maintenance procedures.

Flushing programs can be reactive (emergency), routine, or systemwide. Reactive flushing is often a response to a customer complaint (Figure 8-1). An example of routine flushing would be regular maintenance of known dead-end mains or other trouble spots. A scheduled, systematic, systemwide flushing program can result in long-term water quality improvements.

Extensive flushing programs can become expensive and time consuming. However, the cost of emergency repairs or reactive flushing over a large area can also become costly. Each system must evaluate the scope and cost of its flushing programs. Some costs can be saved by coordinating the scheduled flushing with other maintenance activities (like fire hydrant maintenance or valve testing). Most systems will incorporate all three types of flushing programs in their overall system maintenance plan.

Flushing Procedures

Flushing involves opening a hydrant located near the problem area. The hydrant should be kept open as long as needed to flush out the sediment; this may require up to three water changes. The amount of water needed can be estimated by calculating the volume in the pipeline and the flow from the hydrant. Distribution system water quality data can help operators determine how frequently to flush different areas of the system. The duration of flushing may be determined mathematically, by calculating the amount of time required to flush the appropriate volume of water. The flushing length can also be verified through basic water quality testing, such as chlorine residual and turbidity, performed during flushing. Some systems find that dead-end mains must be flushed as often as weekly to avoid

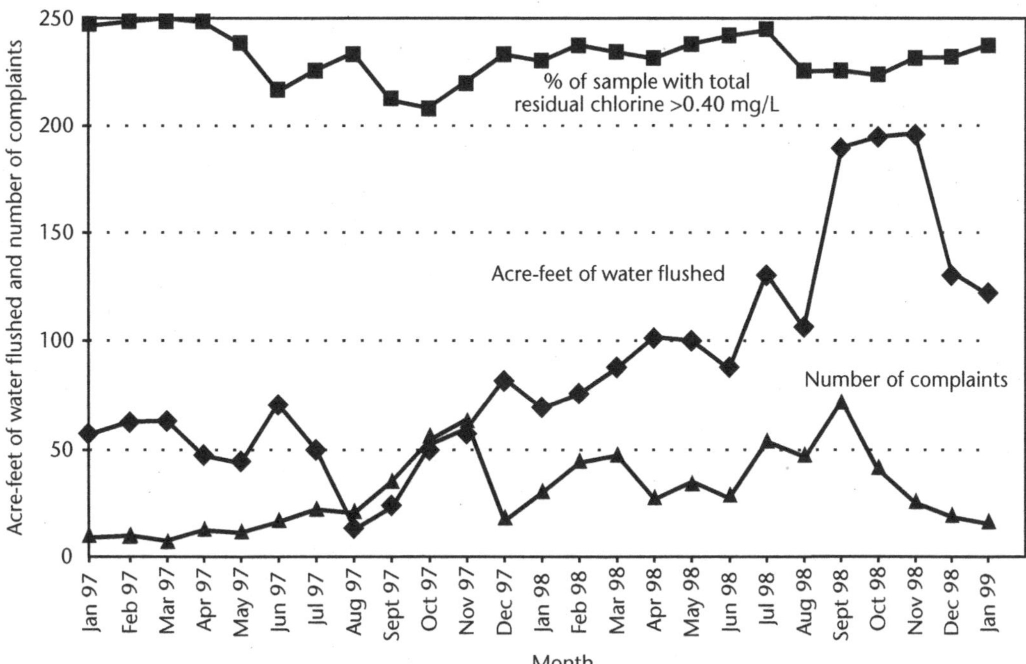

Figure 8-1 Southern California Water Company flushing performance measurement data analysis

Courtesy of Southern California Water Company.

customer complaints of rusty water. It is not normal practice for larger utilities to flush the entire system, but problem areas are often flushed in response to customer complaints. Large areas of the system may receive systematic flushing on a rotational basis, thus resulting in an entire system flushing over a number of years.

The following points should be considered when developing a flushing program:

- A map of the system and past experience should be used to plan (Figure 8-2) the flushing schedule. Emphasis should be placed on areas where there has been a high incidence of customer complaints.
- If the complete system is to be flushed, the process should begin in the area of the well or treatment plant and then progress outward. If only one area is to be flushed, the process should be started at the point closest to the source.
- Flushing the system late at night will achieve greater flows through the line and cause fewer customer complaints. In addition, fewer customers will see how bad the water sometimes looks coming out of the system.
- If possible, media announcements should be made in advance to explain the flushing schedule. This will alert customers that there may be a temporary condition of discolored water. Posting signs in the area to be flushed may also assist in customer relations.
- After flushing is completed, customers should be advised in advance to run enough water (cold water from the bathtub is best) to flush their service lines before using the water. They should also be told that the water is safe to drink but warned not to wash clothes while the water is turbid because it may discolor white clothing.
- The work of flushing crews should be coordinated to avoid flushing too many hydrants at once. If too many hydrants are opened at the same time, a reduced pressure situation could be created, which could increase the chances of backflow through any existing cross-connections.

- Before flushing begins, the area in which the flushed water will drain should be inspected to ensure that the water will not flow into basements, excavations, or buildings. In addition, storm drains should be checked to make sure they are open. Diffusers (e.g., water retention dams) may be needed to reduce erosion.
- Local regulations may require dechlorination of flushed water. Devices that are used to control erosion may also be useful to ensure adequate dechlorination.
- The flow required for effective flushing usually is at least 2.5 ft/sec or 220 gpm in a 6-in. main (0.76 m/sec or 833 L/min in a 150-mm main). A flow of 3.5 ft/sec or 310 gpm (1.07 m/sec or 1,173 L/min) is considered better. Care should be taken to avoid an excessively high velocity, which could cause scouring of pipelines (flow should generally be limited to no more that 10–12 ft/sec [3.1–3.7 m/sec]).
- It may be necessary to turn on additional wells or booster station pumps during flushing to ensure that adequate water quantity and pressure are available.
- Hydrants used for flushing must be opened fully; the hydrant valve is not intended for throttling flow. (Hydrants may be fitted with a gate valve on one of the connections if a low flow is desired.)
- Hydrant valves must be opened and closed slowly to prevent water hammer.
- A nonrigid diffuser, screen, length of fire hose, or other means of reducing the force of the water stream is recommended, especially in unpaved areas (Figure 8-3). Flushing should be stopped if the water is damaging a road or parkway. If damage has already been done, it should be marked with a lighted barricade. The location should be recorded so that the damage can be repaired as soon as possible.
- Flowing hydrants should not be left unattended.
- Flushing should be continued no longer than necessary to remove sediment and ensure that the hydrant is operating properly. Monitoring water quality parameters (e.g., chlorine residual, turbidity, pH) may be useful in determining when flushing has achieved the desired results and can be terminated. In some cases, flushing at a low velocity (using the proper throttling valve) may be adequate for the intended purpose.
- When hydrant use is completed, the hydrant should be checked to ensure that the barrel drains properly. An operator can usually determine this either by placing a hand over one nozzle to feel if a slight vacuum is formed or by listening for air being drawn in when an outlet cap is screwed on loosely.
- Hydrants that have had their drains plugged must be pumped out if there is a possibility of freezing.
- Nozzle caps should be tightened to the point that unauthorized persons cannot remove them by hand.
- The code number of each hydrant flushed, the length of time flushed, the water condition at the start and end of flushing, and other special concerns should be recorded.
- Water quality tests should be conducted to document the results of the flushing. Total/free chlorine residual, temperature, bacteria, iron, color, and odor are examples.
- Hydrants found to be defective should be noted, flagged if inoperative (notify fire department if out of service), and reported for immediate repair.

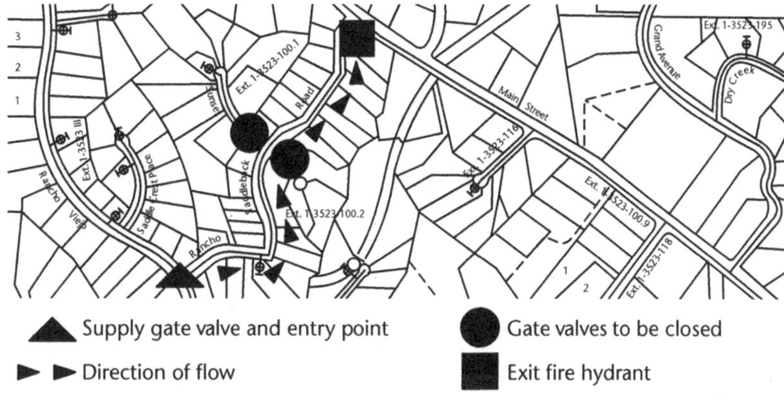

Figure 8-2 Flushing plan map

Figure 8-3 Diffuser used to reduce force of water from hydrant during flushing

Courtesy of Pollardwater.com.

Directional Flushing

This procedure involves a systematic approach to direct the flow, at the desired velocity, from a clean source to the area to be flushed. It is necessary to map the locations of valves that will need to be operated to direct the flow before beginning this procedure. A hydraulic model of the system can be very helpful in planning directional flushing. Proper valve operation is key to this approach; therefore, it is common to conduct valve inspections and maintenance when performing this procedure. Although time is needed to plan the flushing strategy for directional flushing, some utilities have found that this process is actually less expensive than the traditional (open hydrants without operating valves) procedure because water quality is maintained and there are fewer instances of needing to return to repeat the procedure.

Water Main Cleaning

Pipelines that exhibit flow restrictions and/or water quality problems should be cleaned. Utilities should first try to clean the mains by flushing. If flushing proves inadequate, air purging or cleaning devices such as swabs or pigs may need to be used. In addition to removing objectionable material from a main, the cleaning operation can increase the flow rate through the pipe.

Mechanical cleaning may be necessary in areas where excessive tuberculation and deposits are found in older cast-iron pipes or where iron bacteria and slime growth are a severe problem. In some cases, removing encrustation or tuberculation may cause leaks that must be repaired. The cleaning process is not always a permanent solution to dirty water problems. Unless the cleaned pipe is lined or the corrosiveness of the water reduced, the condition will probably come back. However, experience has shown that, in many cases, leaving just a thin coating of iron oxide on the smooth interior of the pipe wall delays the occurrence of red water and the regrowth of encrustation.

Main-Cleaning Preparations

Thorough planning must precede a pipe-cleaning operation. The section of main or system to be cleaned should first be mapped. The order of work, the water source, the entry and exit points, and the disposal method for the flushed water should all be determined. The vehicles to be used, the size of the crew, and the necessary equipment and materials can then be listed and arrangements made for them to be available.

Before cleaning begins, valves and hydrants should be checked to ensure they are operable. Customers should be notified concerning the date and time the system will be out of service. Temporary water service should be arranged for any customers who must have water for medical reasons.

Other utilities and agencies that will be affected by the planned operation must also be notified, including police and fire departments. Regulatory agencies should be consulted concerning any special requirements, and any necessary safety procedures should be planned.

Before the utility work crew flushes or cleans any main, a way to control pressure surges should be provided. The sudden stopping of flow if a line valve is operated too rapidly or if a pig suddenly stops moving can cause a large pressure surge. Such surges can raise system pressure 20–60 psi (138–414 kPa) for each foot-per-second (meter-per-second) of velocity change, which can destroy water mains and appurtenances.

Air Purging

In the air-purging process, air mixed with water is used to clean mains. Generally, small mains, up to 4 in. (100 mm) in diameter, are cleaned using this procedure; however, some utilities have reported using this process for larger mains. Before the procedure is performed, all services must be shut off. Air from a compressor is then forced into the upstream end of a main after the blowoff valve is opened at the downstream end. Spurts of the air–water mixture remove all but the toughest scale.

Swabbing

The swabs used in pipe-cleaning operations are polyurethane foam plugs. These plugs are somewhat larger than the inside diameter of the pipe to be cleaned, and they are forced through the pipe by water pressure. Swabs can remove slime, soft scale, and loose sediment without the need for high-velocity water flow. However, they will not significantly remove hardened tuberculation. Swabs wear out quickly in heavily encrusted mains and must be frequently replaced.

Swabs can be purchased commercially either in specific sizes or in bulk (bulk swabs can then be cut to size). The swab material is available in soft and hard grades. Soft swabs are typically used for mains that have an undetermined cross section or condition. They are also used in mains where the reduction in pipe diameter is expected to be 50 percent or more. They may also be used in mains where there is severe encrustation but where a pig cannot be used because of bends or partial obstructions in the pipe. Hard swabs are commonly used in newer mains, in mains that have minor reductions in diameter, and in mains where deposits need continuous hard-swab pressure.

air purging
A procedure to clean mains less than 4 in. (100 mm) in diameter, in which air from a compressor is mixed with the water and flushed through the main.

swab
Polyurethane foam plug, similar to a pig but more flexible and less durable.

An experienced crew can swab up to several thousand feet of main per day if the operation is planned properly. Swabbing procedures vary with each job, but a typical procedure is as follows:

1. Notify the public and shut off the system.
2. Install the necessary equipment at the entry and exit points for launching and retrieving swabs.
3. Isolate the water main or portion of the system to be cleaned. Be sure all valves on the main to be cleaned are open.
4. Open the valve on the upstream water supply to launch the swab and to control speed.
5. Operate swabs at the speed recommended by the manufacturer. If they travel too fast, they remove less material and wear out more rapidly.
6. Estimate the flow rate by using a Pitot gauge at the exit or a meter on the inlet supply.
7. Note the entry time and estimate the time of exit. If the travel time is too long, the swab may have become stuck. Reverse the flow and time the return to determine the location of blockage.
8. Perform enough swab runs that flushing water becomes clear within 1 minute following the run.
9. Account for all the swabs to make sure none were left in a main. A typical cleaning operation may take from 10 to 20 swabs.
10. Do a final flush until the water is clear of swab particles.

Pigging

Pipe-cleaning pigs are stiff, bullet-shaped foam plugs that are forced through a main by water pressure. They are similar to swabs but are harder, less flexible, and more durable. These differences allow the pigs to remove harder encrustations. However, pigs have more limited flexibility, which somewhat reduces their ability to change direction at fittings and at points where there are significant changes in pipe cross section.

Pigs are purchased commercially in various sizes, densities, grades of flexibility, and external roughness. A number of different types of pigs are available for use in system cleaning. Special pigs can be made for most situations. The ones most commonly used are classified as bare pigs, cleaning pigs, and scraping pigs.

Bare pigs, made of high-density foam, are usually the first pigs sent through a tuberculated main to determine the inside diameter of the obstruction. *Cleaning pigs* have a tough coat of polyurethane synthetic rubber applied in a crisscross pattern. When sent through a main, a cleaning pig removes most types of encrustation and growths. A bare pig made of low-density foam or a swab is sometimes sent behind an undersized cleaning pig to maintain the seal. *Scraping pigs* have spirals of silicon carbide or flame-hardened steel-wire brushes. These pigs are used to remove harder encrustations and tuberculation. Cleaning or scraping pigs of increasing size may be sent through a main to remove layers of encrustation gradually.

Figure 8-4 shows three methods of launching pigs with permanent or portable launchers. During a launch from a fire hydrant base, an external source of water is required to force the swab or pig down the hydrant and into the main. This water is supplied either by a fire hose connected to a hydrant not in the isolated section or by a small, high-pressure pump with an independent water supply.

pig
Bullet-shaped polyurethane foam plug, often with a tough, abrasive external coating, used to clean pipelines. Forced through the pipeline by water pressure.

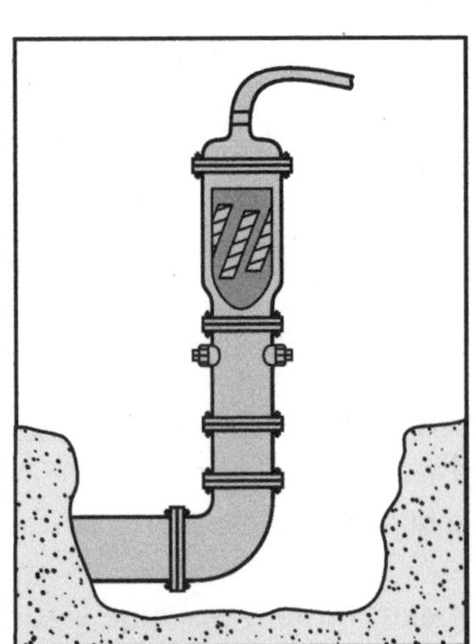

Special pig launcher, attached to fire hydrant with valve removed

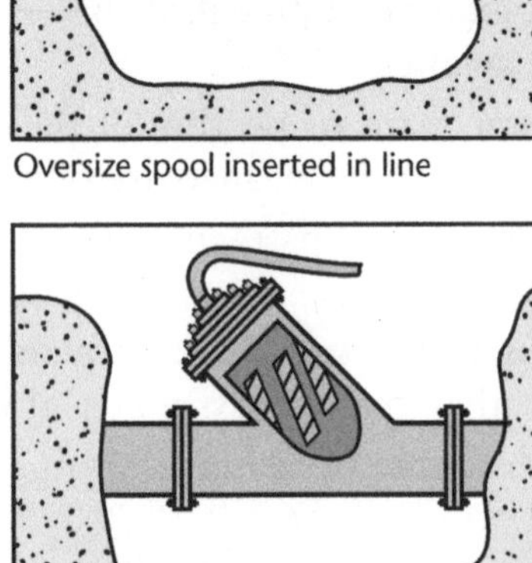

Oversize spool inserted in line

Y-section inserted in line

Figure 8-4 Launching methods for pipe-cleaning pigs
Courtesy of Girard Industries.

After a pig has been forced into the main, the hydrant branch can be isolated. The pig will then be pushed along the main by water from the distribution system. Operators may be able to launch a pig into larger mains by removing the gates from a gate valve and inserting the pig into the main through the body of the valve.

Pigging procedures vary with the anticipated condition in the pipeline, the location of the section that needs cleaning, and the type of pig to be used. The utility should determine the procedure to be used at each location independently, especially for larger pipe sizes. The first time pigs are used, the utility's crew should work with someone experienced in using the equipment and performing the cleaning operation. Assistance is available from pipeline-cleaning firms and manufacturers of cleaning devices. A general procedure for using pigs to clean a line is as follows:

1. Notify the public and shut off the system.
2. Install the necessary equipment at the entry and exit points for launching and retrieving pigs. In many cases, it will be necessary to cut into the main to install the equipment.
3. Isolate the water main to be cleaned. Be sure all gate valves on the main to be cleaned are open.
4. Make provisions to control surges.
5. Open the upstream water supply to launch the pig.
6. Time the passage of the pig in order to gauge the valve setting required to achieve the desired speed.
7. Control the speed of the pig with the downstream hydrant or blowoff valve. Typical speeds are 1–5 ft/sec (0.3–1.5 m/sec). If pigs travel too fast, they remove less material and wear out more rapidly.
8. Avoid sudden changes in speed or stops in pig movement, which will cause destructive surges in water pressure.
9. Run the final flush until the water turns clear.

WATCH THE VIDEO
Flushing (www.awwa.org/wsovideoclips)

Metal Scrapers

Cleaning units that are forced through a main by water pressure are also available in designs that have metal scrapers. They consist of a series of body sections on which high-carbon spring-steel blades of various shapes are mounted to provide scraping and polishing. The sections are free to rotate and are pushed through the main by water pressure acting against pusher cones.

Power-Driven Cleaning

When deposits in a water main are particularly thick or dense, mechanical cleaning with a power drive unit is needed. In this process, a rod similar to a sewer rod is used to pull a cutter through the main to dislodge the built-up material. This type of work must generally be performed by specialized contracting companies.

Final Cleaning Procedures

After a water main has been cleaned, it should be flushed until the water runs clear. The main should also be chlorinated before it is returned to service. All valves should be checked to make sure they have been left in the open position and that all customer water services have been reactivated.

Before and after the cleaning is performed, a flow test should be conducted on mains that are to be cleaned. The results of the test after cleaning should indicate if the procedure was successful or if further cleaning is necessary.

Chlorine Treatment

When the carrying capacity is reduced by slime growths in pipelines, chlorine may be used to solve the problem. The section of the system affected is usually treated with the slug method, whereby a "slug" dose of chlorine is used to kill the bacteria causing the problem (usually 25 mg/L for 24 hours; specific procedures are detailed in AWWA Standard C651, most recent edition). This slug should be followed by thorough flushing to get the slime and chlorine out of the system. The precautions discussed previously concerning flushing are even more important in this situation, *especially* those that relate to informing customers to flush their service lines thoroughly before using the water.

A second dose of chlorine may be needed to complete the job. A disinfectant residual should be maintained throughout the system after the treatment is done to prevent the bacterial slime from regrowing.

Lining Water Mains

Cleaning can usually restore interior pipe surfaces to a condition close to that of a newly laid main. However, experience has shown that cleaning iron pipe without lining it is only a temporary solution. If the water quality remains unchanged, tuberculation will occur again at an even faster rate after cleaning. The flow coefficient will decline back to its previous level. For this reason, cleaning alone does not accomplish much.

Cement-Mortar Lining

After cleaning is finished, a thin layer of cement mortar can be applied to the pipe walls to line them in place. This not only prevents interior surface deterioration from recurring but also results in improved water quality, volume, and pressure to customers.

slug method
A method of disinfecting new or repaired water mains in which a high dosage of chlorine is added to a portion of the water used to fill the pipe. This slug of water is allowed to pass through the entire length of pipe being disinfected.

The cost of mortar lining in place depends on the following:

- Pipe diameter and length
- Condition of the pipe
- Layout and profile of the line
- Number of bends
- Locations and types of valves
- The need to provide temporary bypass lines
- Type and depth of soil cover
- Accessibility
- Traffic conditions

The longer the length of pipe that can be lined in one operation, the greater the production rate and the lower the cost per unit length. As illustrated in Figure 8-5, small-diameter pipe is lined by remote-controlled equipment that is pulled through the main. Large-diameter pipe can be lined using equipment that allows a worker to enter the pipe to control the operation.

When distribution mains are lined, temporary water service must be provided to all customers until the operation is completed. Water utilities usually provide this service by running a 2-in. (50-mm) pipe on the surface over the main. All services must be connected to this pipe through meter pits or to building sill cocks. After the lining has been placed, each service must be dug up and the mortar removed from corporation stops before service can be restored to customers.

Slip Lining

Another method of lining cleaned water mains in place involves slip lining with a high-density polyethylene pipe. The plastic pipe is lightweight and flexible and can be either pulled or pushed into the existing main from access points cut into the system. Valves, tees, and services must be recut, the same as with cement-mortar lining.

Booster Disinfection

Most systems perform disinfection only at the water treatment plant or wellhead. The dosage applied is adequate to provide a measurable residual throughout the distribution system. In some cases, this is not possible without adding a prohibitive amount at the source. Booster disinfection (redisinfection) is needed in this

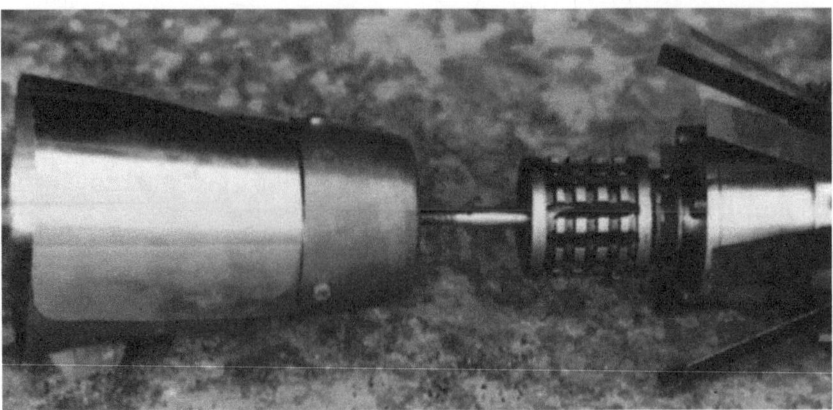

Figure 8-5 A cement-mortar lining machine for use in small-diameter pipe

instance. Some utilities have employed booster disinfection as a strategy to maintain a more uniform disinfection residual at a lower level, reduce DBPs, and lower total disinfection cost (in some cases).

Where chloramines are used to provide the disinfectant residual, booster disinfection requires special consideration. If free chlorine is applied for booster disinfection, it is possible for free chlorine from this process to blend with chloramine residual in the system in an uncontrolled manner. The potential problems of blending are related to the breakpoint curve (Figure 8-6) for free and combined chlorine. Various resultant concentrations of mono-, di-, and free chlorine are possible. This may lead to undesirable tastes and odors or possible reduction of chlorine residual in some areas. Some utilities have dealt with this situation by measuring the ammonia residual remaining in the water at the point of free chlorine addition. They have then added the correct ratio of free chlorine to reform predominately monochloramine. In this case, blending with residual chloramine is not an issue. Other utilities have found that free chlorine booster disinfection in a chloraminated supply does not cause problems. The best strategy for each utility is dependent on many factors, some of which are unique to the site. Therefore, each situation should be fully evaluated before implementing any disinfection strategy.

Customer Complaints

Many of the operational practices and system design features that are described in this chapter and throughout the book are directed toward promoting customer satisfaction. A valid customer complaint is evidence that 100 percent satisfaction has not been achieved. In this way, customer complaints indicate a failure by the water supplier to *prevent* complaints.

No distribution system is completely free of situations that occasionally create water quality problems and thus generate customer complaints. Water quality complaints should be tracked separately from other customer issues. The information from a complaint investigation can be useful in identifying the source of a deteriorating water quality condition. Persistent complaints in one area may lead to a change in operational practices or system alterations to permanently address the problem. The utility operator should establish a goal to reduce customer complaints to improve customer satisfaction.

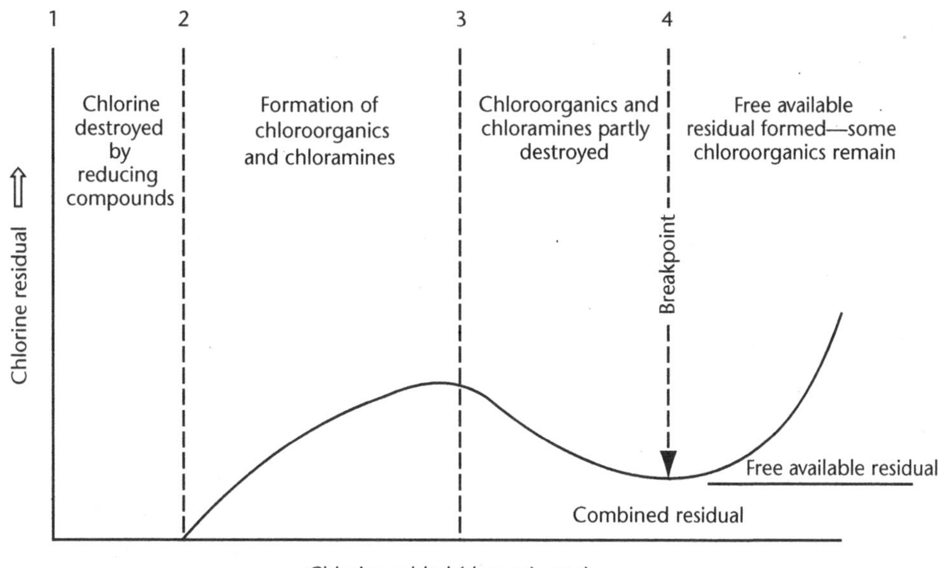

Figure 8-6 Breakpoint chlorination curve

Source Water Blending

When a utility has access to more than one source of water, a detailed blending analysis should be conducted. Blending can have beneficial or detrimental consequences on distribution system water quality and pipe materials. Conservative parameters such as iron, phosphate, fluoride, and manganese may be estimated using blending models that are based on concentration and flow. Parameters that decay or grow under certain circumstances may need more complex models to predict system concentrations.

Blending may result in conditions that affect the stability of pipeline materials and the films that adhere to internal surfaces. For example, changes in water chemistry can reduce or increase the rate of corrosion. Alternatively, microbiological counts have increased in pipelines subject to blended supplies. Many of these situations can be predicted and adverse consequences can be avoided by planning and modeling blending practices. It may be necessary to avoid blending by isolating areas of the distribution system with valves or by using pressure zones (this practice may create dead-end areas that could develop water quality problems). Care should be taken to notify customers if changes in water quality are anticipated.

Source Water Treatment

The quality of water entering the distribution system can significantly impact the receiving system and customers' satisfaction with their drinking water. The system operator's goal is to maintain water quality at the point of entry to the customers' service (the extent of the system operator's responsibility may be defined by local regulations). This goal can be achieved only if the source water treatment is optimized to match the system conditions to prevent water quality changes from occurring.

Several major treatment practices (discussed in more detail in the *WSO: Water Treatment* certification books) can greatly aid system operators in their quest to maintain water quality, including the following:

- Stabilizing pH
- Controlling corrosion
- Optimizing primary disinfection
- Optimizing turbidity removal
- Reducing organic compounds
- Minimizing iron and manganese concentrations

Most water quality changes that can be attributed to source water treatment can be addressed with one or more of these practices.

The pH of the water affects many common reactions. Therefore, pH stability is critical to maintaining water quality throughout the distribution system. Corrosion of system components, the rate of DBP formation, disinfectant decay, and tastes and odors can all be influenced by pH. The buffer intensity is a measure of the resistance to a change in pH. This measure is better than alkalinity for determining the pH stability of a given water. The effect of pH (and alkalinity) on buffering intensity for the carbonate equilibrium system is shown in Figure 8-7. The minimum buffering intensity occurs at pH 8.3 for this system, which is why it is very difficult to maintain a stable pH at this value. The buffering intensity can be increased by adding carbonate and raising the dissolved inorganic carbon (DIC) level.

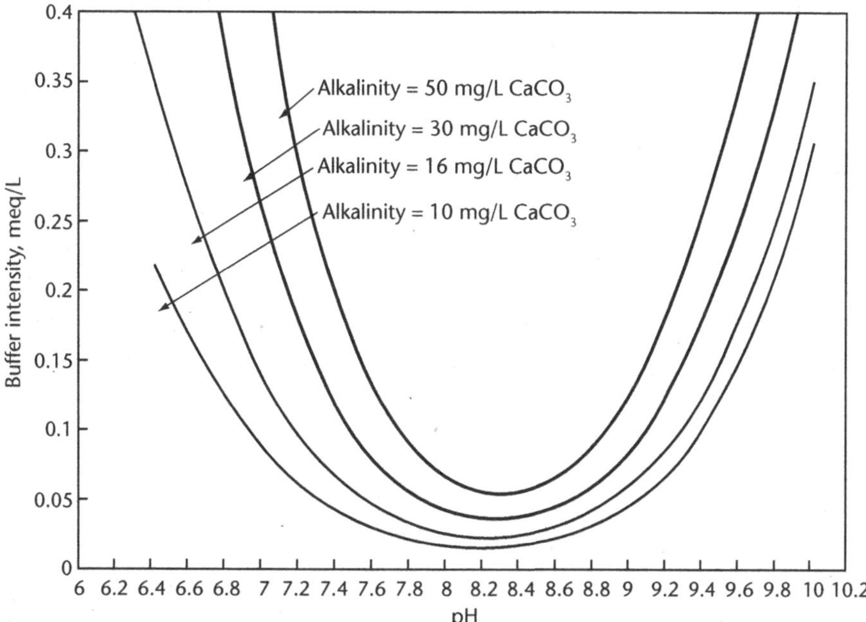

Figure 8-7 Theoretical buffer intensity versus pH

Many natural and treated waters are corrosive and will dissolve some pipe and plumbing materials. Corrosive waters can deteriorate both the distribution system piping and domestic plumbing systems. This deterioration (or **corrosion**) can, in turn, cause the quality of the water delivered to the customer to be significantly degraded. Corrosion can also reduce flow capacity and shorten the life span of the distribution system.

Figure 8-8 illustrates a corrosion cell and shows how corrosion can cause pitting and the formation of rough tubercles. Even slight corrosion and tuberculation can increase the roughness of a pipe's surface, thus significantly reducing the pipe's carrying capacity. Major corrosion can also weaken pipes. Corrosion products may break off of mains, causing clogged services and customer complaints of rusty water.

The principal concern with internal corrosion in the past was its effect on unlined cast-iron or steel mains, steel tanks, and other metal surfaces in the distribution system. As a result, unlined cast-iron and steel pipe are no longer installed, but there are still hundreds of miles of old unlined pipe in use in older systems. Corrosion of the interior of steel tanks is generally controlled by a protective coating or by cathodic protection.

Internal pipeline corrosion control is usually implemented to reduce lead, copper, and iron leaching into treated water. Strategies used for this purpose involve pH and alkalinity adjustment, the use of corrosion inhibitors, and calcium adjustment. Lead control in the pH range of 6–9 generally requires the DIC to be greater than 2 mg C/L (2 mg carbon/liter). Carbon dioxide, soda ash, and sodium bicarbonate may be used to adjust the DIC. Orthophosphate may be used as a corrosion inhibitor in the pH range of 7.4–7.8 at a typical residual of 1–5 mg PO_4^{3-}/L. Copper and iron corrosion are reduced by using orthophosphate or by adjusting pH and DIC.

Effective primary disinfection protects the distribution system from pathogens that may be present in the source water. The USEPA specifies the disinfection requirements for drinking water systems. Refer to Chapter 1 for more information on current water regulations.

corrosion
The gradual deterioration or destruction of a substance or material by chemical action, frequently induced by electrochemical processes. The action proceeds inward from the surface.

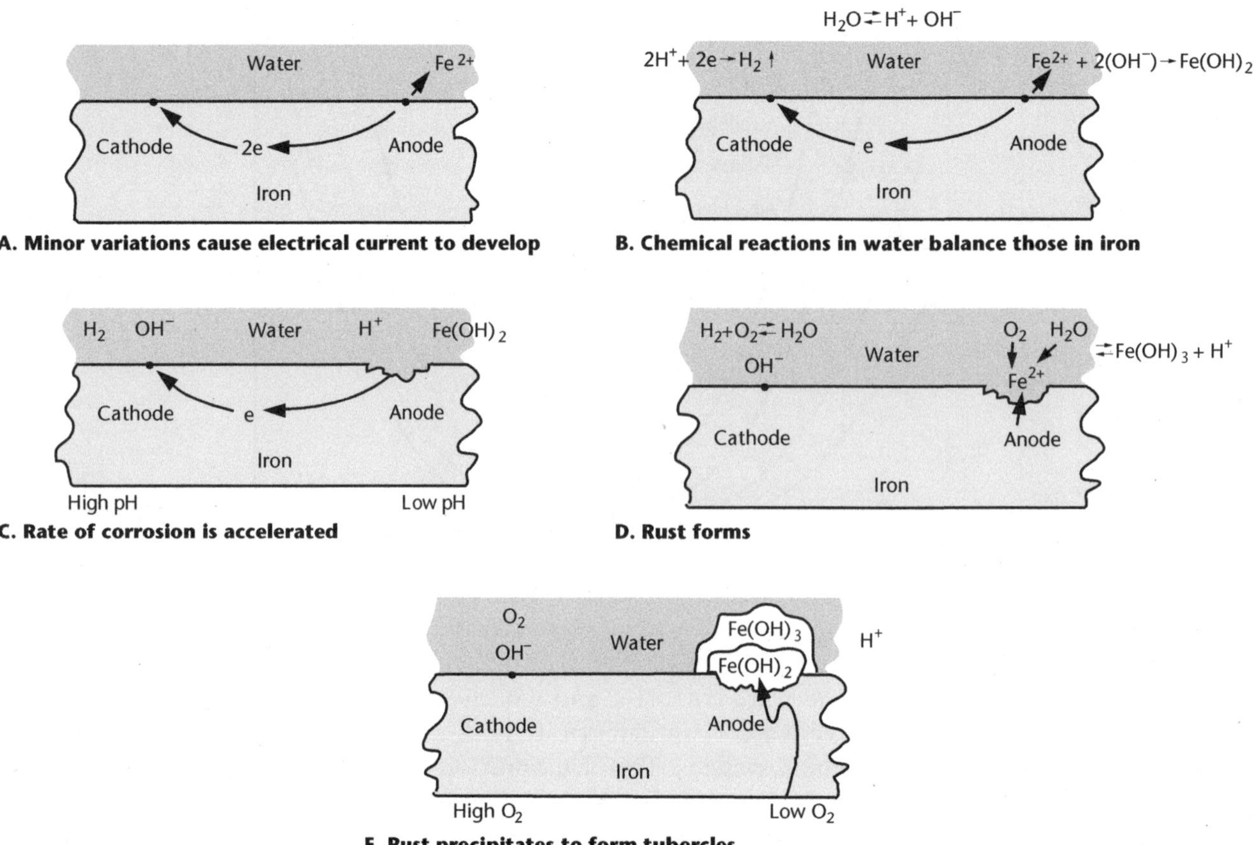

Figure 8-8 Chemical and electrical reactions in corrosion cell

Most systems provide primary disinfection with free chlorine or ozone. However, monochloramine, chlorine dioxide, and ultraviolet light may also be used under specific conditions.

Water treatment plants are designed to remove particles. Turbidity is a measure of particles that can scatter light. Lower-turbidity water can be disinfected more efficiently and disinfectant demand in the distribution system may be decreased. Water treatment optimization programs such as the Partnership for Safe Water recommend the lowest turbidity possible. This procedure may lower the risk of possible exposure to pathogens such as *Cryptosporidium*.

The removal of organic compounds by activated carbon filtration, ozone biofiltration, membrane filtration, or enhanced coagulation may reduce the occurrence of bacteria in the distribution system. Assimilable organic carbon and natural organic matter (NOM) have been shown to support regrowth of bacteria. Chlorination of NOM has also led to increases in DBPs. Both the regrowth of bacteria and the development of DBPs are problems that may be difficult to overcome in the distribution system. Source treatment is therefore recommended if these conditions are encountered.

High levels of iron and/or manganese can cause discoloration of water, fixtures, and laundry and adverse taste. Removal of these elements from water may lead to improved customer satisfaction. Lime or caustic soda softening or oxidation/filtration are used for removal. Many plants that previously used free chlorine for oxidation have changed to potassium permanganate due to concerns over the production of chlorinated DBPs. A common treatment process is oxidation with potassium permanganate and greensand filtration (a specialized filtration media suited for manganese removal).

Seasonal Considerations

Numerous water quality conditions are affected by seasonal changes either in water use or availability. Source water quality often changes seasonally. As water use decreases, hydraulic detention time may increase, unless the system operator takes action to prevent this consequence (e.g., removing excess storage from service in the winter). In many cases, this situation occurs in the winter when the water temperature is lower. This relationship may be beneficial because many adverse water quality conditions are aggravated by increased temperature.

Some utilities may augment water supplies during high-demand periods. Uncontrolled blending of waters of different water qualities can occur in the distribution system. The operator should consider the consequences of this practice and take steps to maintain uniform water quality at least in confined areas of the system (e.g., by creating temporary pressure zones). Flow reversal and pressure surges should also be avoided when introducing supplemental water sources.

Nitrification is the process by which ammonia-oxidizing bacteria convert ammonia to nitrite and nitrate. Most nitrification episodes are associated with warmer water conditions and occur in locations of low water turnover. Overdosing ammonia in chloraminated systems is a common cause of nitrification. Chloraminated systems should therefore monitor free ammonia as a control parameter. Systems may employ seasonal operations such as practicing periodic free chlorination, practicing breakpoint chlorination, and increasing the turnover rate in storage facilities to prevent nitrification. Symptoms of nitrification are loss of chlorine residual and increases in nitrite and/or nitrate concentration. The loss of chlorine residual may also lead to the occurrence of coliform bacteria in the system, thus leading to the possibility of regulatory noncompliance. Minimizing detention time and maintaining a chlorine/chloramine residual are key steps in preventing nitrification. The system operator should assess storage facilities in particular for the possibility of nitrification in isolated sections of the facility (the entire water volume may not be affected). Spot treatment or cleaning may be required to eliminate this condition. For more information, consult AWWA Manual M56 *Nitrification Prevention and Control in Drinking Water*.

Pressure Requirements

An important operational requirement for any distribution systems is to maintain a continuous positive pressure at all locations. **Backflow** or **backsiphonage** from **cross-connections** may occur when there is negative or zero pressure in pipelines. Contaminants may be drawn into the system through leaking pipes, submerged air-and-vacuum relief valves, blowoff valves, or faulty check valves. Standards for the minimum pressure in the distribution system vary. However, 20 psi (138 kPa) is the lowest minimum pressure listed in many existing US state standards and water system guidelines (e.g., the Ten States Standards and National Fire Protection Association standards).

Excessively high system pressures can also cause water quality problems. Higher incidence of main breaks and service repairs can increase the water quality disturbances caused by these procedures. The effects of water hammer are increased by higher pressure. The resultant hydraulic surges can cause material on pipe walls to become dislodged, thus degrading water quality.

Cross-Connection Control

Backflow from nonpotable water sources that are cross connected to the potable system can cause serious water contamination incidents. Both backsiphonage and **backpressure** can be the mechanism for the backflow. Where cross-connections

backflow
A hydraulic condition, caused by a difference in pressures, in which nonpotable water or other fluids flow into a potable water system.

backsiphonage
A condition in which the pressure in the distribution system is less than atmospheric pressure, which allows contamination to enter a water system through a cross-connection.

cross-connection
Any arrangement of pipes, fittings, fixtures, or devices that connects a nonpotable system to a potable water system.

backpressure
A condition in which a pump, boiler, or other equipment produces a pressure greater than the water supply pressure.

exist, backflow may be prevented by the use of approved control measures such as the following:

- Air gap
- Reduced pressure zone backflow-prevention device
- Double check valves
- Pressure vacuum breaker
- Atmospheric vacuum breaker

The application and acceptability of each measure depends on the degree of hazard and the conditions of the potential backflow situation. In many areas, local regulations define the appropriate measure.

All water systems should implement cross-connection programs. The program should do the following:

- Define authority and responsibility
- Require systemwide inspection and testing of all prevention devices
- Define device maintenance and record requirements
- Educate all parties

Emergency Operations

Water quality problems can cause an emergency situation, and emergencies can cause water quality problems. Natural disasters, main breaks, and power outages are examples of emergency situations that can degrade water quality. Likewise, earthquakes, floods, and violent storms can lead to contaminated water incidents.

Dealing with these situations at the time they occur without sufficient planning can cause inefficient and sometimes incorrect responses. Utilities should develop an emergency contingency plan that includes the following:

- An emergency response strategy
- Interconnections with other utility supplies
- Sufficient storage for emergency operations
- A plan for natural disasters
- A conservation plan to help with insufficient supplies

The plan should address the recovery period that leads the utility back to normal operations. Operators should be familiar with the plan and should participate in periodic simulation drills to test and refine the plan.

Energy Management

The focus of most energy management strategies is to operate the system to minimize power expense. The effect on water quality should also be included when considering options regarding energy usage. Strategies to reduce costs of energy use include treatment plant scheduling, pump scheduling, water demand forecasting, and maintenance scheduling. Water quality can be affected when employing these strategies because hydraulic detention time can be changed and water velocity or direction can be influenced.

Redundant Power Supply

Improving power supply reliability (redundant power supply lines, portable generators, or permanent emergency generators) can reduce the frequency of system shutdowns. Pipeline disturbance due to on–off operation is therefore lessened.

Pumping Strategies

Pumping operations should consider velocity in pipelines and the effect of on–off operations and include sequences that effectively increase the turnover rate in storage facilities.

Pipeline Friction Loss

Friction losses are the primary factor affecting power requirements for pumps. Regular flushing, corrosion control, and pipeline cleaning can reduce power requirements and improve water quality at the same time.

Direction and Velocity Control

Water quality can be affected by rapid changes in velocity or the direction of flow in pipelines. Operators should employ procedures that will avoid or reduce the impact of these changes.

Fire Flow Testing

Fire flow requirements are the determining factor in many systems when sizing mains, storage facilities, and pumps. Fire flow tests are necessary to ensure that the system is adequate to provide the necessary flow for firefighting. The flow requirements are specific to each community and are dependent on a number of factors.

One or more fire hydrants are used for these tests. The hydrants are operated from zero to full flow while measuring the pressure at a nearby hydrant. The rapid change in velocity in the main feeding the hydrants can cause water quality problems. Water hammer created from opening and closing hydrants can add to these problems. Customer complaints are often a direct result of fire flow testing. It is best to alert customers that this procedure will be conducted in their area and advise them that they may notice a temporary change in their water quality. The notice should suggest how they can help clear the water following the procedure.

System operators can also take advantage of these tests and coordinate other procedures at the same time. For example, if the tests are performed in a manner consistent with the needs of the routine flushing program (discussed earlier in this chapter), both procedures can be accomplished at the same time. Hydrant valve inspection and maintenance can be conducted. It may also be a good time to inspect the function of air-and-vacuum relief valves and check backflow-prevention devices.

Pump Startup, Shutdown, and Valve Operation

Resuspension of sediment and scouring of scale and other attached material on pipe walls may occur when there is a variation in flow velocity or direction. Water quality can be negatively affected by these operations, resulting in customer complaints. Therefore, operations that cause flow variations should be minimized.

Valves and hydrants should be opened slowly. Pumps should be started with the discharge valve closed. As the motor reaches full speed, open the valve slowly. The reverse procedure will help avoid problems upon pump shutdown. Automatic pump control valves can perform this sequence without operator attention. Variable frequency drives can be used on pumps to slowly change the pump speed upon startup and shutdown. If this is not possible, relief or surge chambers may be installed to absorb the pressure shock from shutdown.

Pump operation may also cause a change in the direction of flow. This procedure can result in many of the same water quality problems as changes in flow velocity. System operators should evaluate alternatives to minimize this condition.

Maintaining Flow and Pressure

The second objective of distribution system operators is to ensure system reliability. This is achieved by following practices and procedures that result in positive pressure throughout the system and the delivery of adequate flow for all intended purposes. This section discusses some of the many concerns in keeping a system in working order.

Distribution System Inspection

The performance of distribution system piping depends on the pipe's ability to resist unfavorable conditions and to operate at or near the capacity and efficiency that existed when the pipe was laid. This performance can be checked using flow measurement, fire flow tests, loss-of-head tests, pressure tests, simultaneous flow and pressure tests, and tests for leakage. These tests are an important part of system maintenance. They should be scheduled as part of the regular operation of the system.

Pressure and Flow Tests

Some state regulatory agencies and local codes require public water utilities to maintain normal pressure in the distribution system between 35 psi (241 kPa) and 100 psi (689 kPa), with a minimum of 20 psi (138 kPa) under fire flow conditions (National Fire Protection Association, 2007).

Many changes can occur in a distribution system to reduce (or occasionally increase) pressure and flow to unacceptable levels. Such changes include the following:

- System expansion that does not provide additional feeder mains
- New water lines installed at elevations higher or lower than the original system
- Additional customer services added to existing mains
- Unintentionally closed or partially closed valves
- Undetected leaks in mains or services
- Changes to water storage tanks
- Reductions in pipe capacity due to corrosion, pitting, tuberculation, sediment deposits, or slime growth

Minimum pressures must be maintained to ensure adequate customer service during peak flow periods or while water is being used to fight a fire. A minimum positive pressure must be maintained in mains to protect against backflow or backsiphonage from cross-connections.

Excessive pressure is objectionable for use by customers and decreases the life of water heaters and other plumbing fixtures. It can also increase the chance of distribution system damage in the event of water hammer. Some plumbing codes (e.g., Uniform Plumbing Code) and state or local regulations require the installation of pressure-reducing valves on high-pressure services.

Checking Pressure Operators can test water distribution system pressure using a pressure gauge connected to either a fire hydrant or a building faucet. Hydrant pressure gauges (Figure 8-9) can be purchased or made from a standard hydrant cap that has been tapped. After the gauge is installed on a hydrant nozzle, the hydrant is opened far enough to ensure that the drain valve has closed. Air is then bled off through the petcock before the reading is taken. The pressure is read in pounds per square inch (psi) or kilopascals (kPa).

If a hydrant is not available, operators may need to check pressure using a sill cock on the exterior of a residence or public building. When this is necessary, the operator should first obtain the permission of the resident or property owner. Water cannot be used in the building while the reading is taken. Pressure gauges made for use on a hose thread faucet are readily available from plumbing supply houses.

Checking Loss of Head Operators can determine loss of head by isolating a section of the distribution system from all branch lines and services. All water entering the test section of pipe must exit the downstream end. As shown in Figure 8-10, pressure is then measured at points along the main or on customer services to determine how much pressure drop is occurring.

Checking Flow Flow readings are usually taken at hydrants. When locations on the distribution system are found to have reduced flow capacity, the cause may be an increase in pipe interior roughness. This problem is often found in old, unlined cast-iron pipe.

Flow tests conducted on a main over a period of years may gradually indicate decreasing flow rates. This problem is probably due to the added friction or

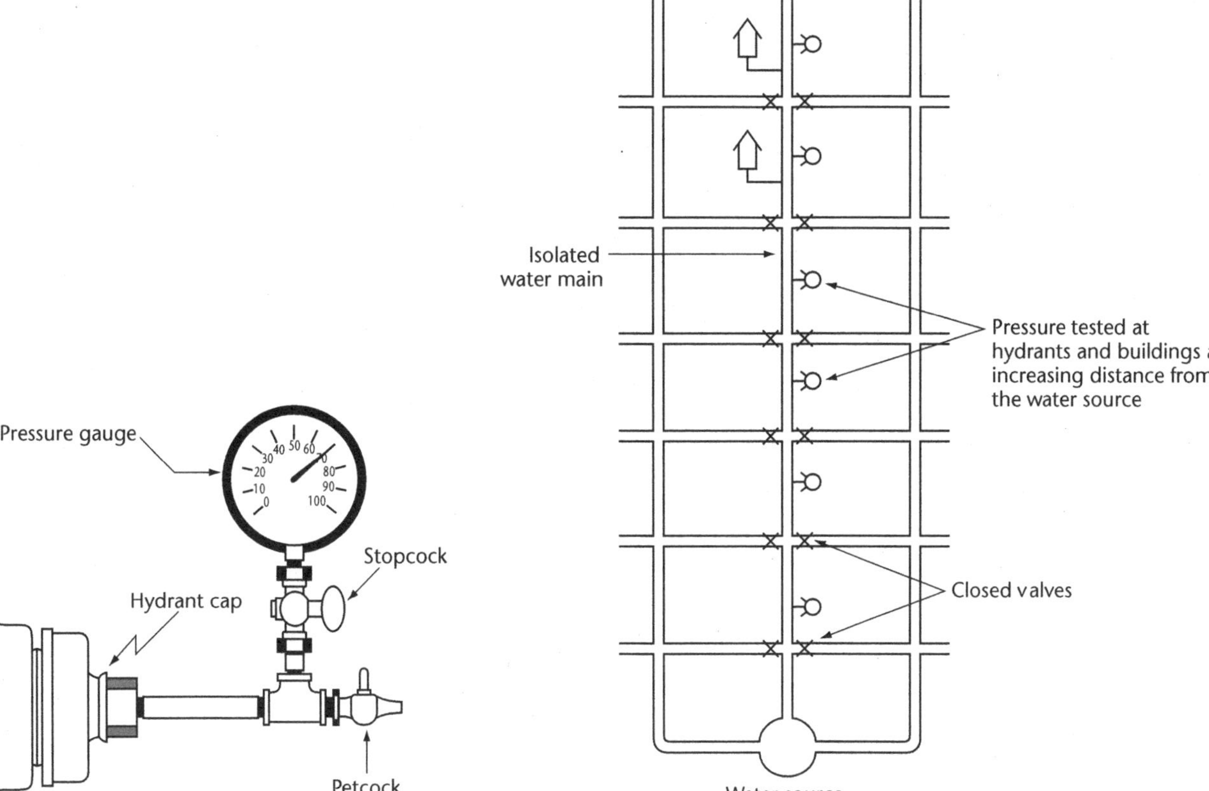

Figure 8-9 Fire-hydrant pressure gauge

Figure 8-10 Locations where head loss is checked in an isolated section of the distribution system

reduced open flow area caused by tuberculation of the pipe interior walls. The calculated flow coefficient (or C value) should be compared with the value for the pipe when it was new. If the flow in a main is reduced to the point where adequate fire flow is no longer available, the utility must either clean the main to restore its capacity or install new feeder mains to reinforce the system.

Routine Inspection

Fire hydrants and distribution system valves should be regularly inspected and operated to determine whether repair is required. Inspection and operation also keep the hydrants and valves "loose" so they will operate freely when needed.

Water utility employees should be specially trained to watch for potential water system problems and notify management of anything suspicious. Meter readers are in a particularly good position to watch for suspected leaks, vandalism, damaged equipment, and unauthorized use. Police, street crews, firefighters, and other municipal employees should also be encouraged to report anything about the water system that appears unusual. The utility may already know about the suspected problem or the situation may not actually be a problem, but the person making the report should nonetheless be thanked for his or her interest. Occasionally, these reports can be very important.

Leak Detection and Repair

Distribution system leaks and the resulting water loss fall into two general categories: emergency leaks and nonemergency leaks.

Emergency leaks require immediate attention. Nonemergency leaks include known leaks that may be repaired when time permits, as well as nonvisible leaks that are located as part of a leak detection survey. Although the repair procedures may be the same for both leak conditions, the method of detection and the sense of urgency involved are often quite different.

Locating Leaks

If all water services are metered and all meters are accurate, the difference between the amount of water pumped to the system and the total water metered to customers is known as unaccounted-for water (nonrevenue water). A large portion of unaccounted-for water is due to the various hydrant uses: firefighting, main flushing, and water leaks. If hydrant use is closely controlled and estimates are made of authorized hydrant use, this amount can be added to the total water metered to customers to create a total of known use. The difference between this figure and the amount of water metered to the system provides an estimate of the amount of leakage in a system.

Large Leaks Unless the ground is very porous, large leaks are usually easy to find. The water either comes to the surface or finds its way into a crack in a sewer pipe or access hole. The starting point for reducing leakage in a system is to urge all police and public works employees to promptly report any unusual puddles or running water. Sewer crews should be asked to report any sewers that seem to have unusually heavy flow.

Small Leaks Unfortunately, small leaks do not always come to the surface. Either they are absorbed into the soil or they flow into sewers, where their flow is too small to be noticed. It can be seen from Table 8-1 that a relatively small leak can pass a surprisingly large amount of water over time. And many small leaks can add up to millions of gallons (or liters) of wasted water.

Table 8-1 Water loss versus pipe leak size

	Water Loss*			
	Per Day		Per Month	
Pipe Leak Size	**gal**	**(L)**	**gal**	**(L)**
•	360	(1,363)	11,160	(42,245)
•	3,096	(11,720)	95,976	(363,309)
●	8,424	(31,888)	261,144	(988,538)
⬤	14,952	(56,599)	463,512	(1,754,584)

*Based on approximately 60-psi (410-kPa) pressure.

All water systems generally repair leaks as soon as possible after the leaks are noticed. However, most systems spend little effort looking for hidden leaks. How much effort a utility should make is principally a function of the availability and cost of water. Systems located in water-short areas should have very active leak detection programs to conserve water. Systems that must pay a high cost to obtain or treat water should also be concerned about leaks because the cost of lost water can be substantial.

WATCH THE VIDEO
Leak Detection (www.awwa.org/wsovideoclips)

Leak Detection Methods

A water utility can control underground leakage only by conducting a painstaking survey of the entire system. The two basic methods are listening surveys and a combination of listening surveys and flow rate measurements, sometimes referred to as water audits.

Listening Surveys An acoustic listening device can be systematically used to locate leaks. It can detect the sound waves created by escaping water. Sound waves are picked up by sensitive instruments and amplified so that a technician can hear them. In the hands of an experienced operator, the instruments can help locate a leak with remarkable accuracy.

The equipment used to detect the sound may be either mechanical or electronic. Two mechanical devices that have been available for many years and are still used today are the aquaphone and the geophone. The aquaphone resembles an old-fashioned telephone receiver with a metal spike protruding where the telephone wire would go. The spike can be placed against a water pipe, against a fire hydrant, or on a valve key that is placed on a valve. It amplifies the sound surprisingly well. The listening end of the geophone looks like a medical stethoscope, and the listening tubes are connected to two diaphragms. When the diaphragms are placed on the ground over a leak, the sound is amplified and the operator gets a stereo effect that aids in determining the direction of the leak.

Powerful and versatile electronic amplifiers are now available (Figure 8-11). The kits generally include equipment for listening for sounds at the ground surface, as well as probes and direct-contact sound-amplification devices. Normal sounds such as wind and traffic will be amplified along with the leak sounds. For this reason, adjustable filters are provided in the amplifier to reduce the band of unwanted frequencies and enhance the frequency band of escaping water.

Dead ends, crosses, tees, and partially closed valves may make locating and verifying a leak more difficult. Another problem with the listening method is that

Figure 8-11 Portable laptop leak correlator

not all leaks can be detected by listening on surface appurtenances. In general, the smallest leaks are the loudest. On the other hand, many very large leaks make no detectable noise at all until a metal rod is driven through the ground to the pipe wall to provide a direct connection for the listening device.

Correlator Method Firms specializing in contractual leak detection use much more sophisticated equipment. A leak technician surveys a segment of pipe using transducers, and the data are fed into a computer (called a correlator) for analysis. The computer takes into account the pipe size, type of pipe material, and other factors that affect the speed at which sound travels. It is generally able to accurately pinpoint a leak location. Figure 8-12 illustrates how transducers connected to two valves on a water main provide data to identify a leak location. The procedure for a listening survey can vary from just listening on selected hydrants (a process known as skimming) to completely covering a system by listening on all hydrants, valve stems, and services.

Statistical Noise Analyzer This system is useful in areas where there may be excessive interference. A hydrophone sensor is attached to a hydrant or tap. The system "listens" for a continuous period (e.g., 2 hours), usually at night. The units are usually installed in the system at intervals up to about 1,500 ft (460 m). A noise signature is produced at each point, and the statistical variance indicates a

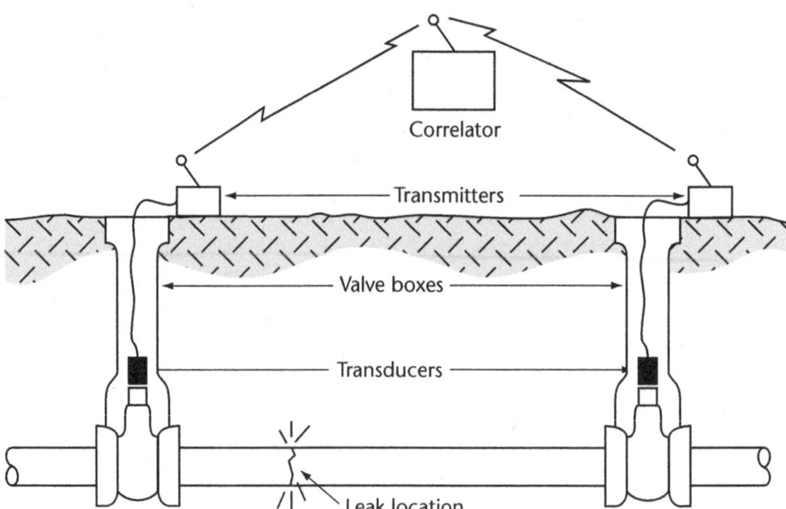

Figure 8-12 Pinpointing a leak from data provided by two transducers

leak and its approximate location. The system has advantages over human listening systems, but the cost of the equipment may be a barrier for some utilities.

Acoustic Leak Detection Procedures Steps in conducting a listening survey usually progress in the following sequence:

1. The system survey is best performed at night, when there is relatively little surface noise and low water flow in mains. Operators use a system map to locate all hydrants in the system. The hydrants are then sounded (checked with a listening device), and those that are noisy are marked on the map.
2. Later, personnel from the transmission and distribution department are consulted for any information that might account for the noisy hydrants.
3. Files are reviewed for any information on abandoned services, previous repairs, and construction that might be the source of a leak.
4. Sewers are checked for possible flow from the potable system. If excessive flow is found in a storm sewer, it may be possible to test a sample for the presence of chemicals that have been added to the treated water, such as fluoride. Finding these chemicals helps confirm the existence of a leak somewhere in the area.
5. Before searching further, the utility should prove that the noise noticed on a hydrant is not from the hydrant valve. Workers can prove this by exercising the valve several times and seeing if the noise stops or the pitch changes.
6. The hydrant auxiliary valve should also be operated to see if doing so makes any difference in the noise.
7. Intersection valves on the mains, the surrounding curb stops, and the customers' outside sill cocks or faucets are then checked for noise level. One of these locations is likely to have significantly louder leakage noise, indicating that the leak is nearby.
8. If the noise seems to be coming from a customer's service line, the service is shut off at the curb stop and the service pressure is relieved by opening an operable faucet. If the noise disappears, the leak is on the customer's side of the service.
9. If the noise is still present after the service is valved off, the leak is on the utility side of the service—most likely on the main near the corporation tap. In this case, the road surface over the line is sounded with a ground microphone. A mark is then placed on the road where the loudest sound is located to indicate the possible leak. The outside limits where the noise is observed to change or disappear are also indicated on the road surface. In this way, a working area can be identified.
10. To further pinpoint the leak with as little damage to the road surface as possible, the following procedure is used. First, a 3-in. (76-mm) hole is drilled through the pavement over the main. Then, a compressed air supply is used to blow a hole down through the subsoil until the main is reached. If the leaking area has been located, water will usually come to the surface through the probe hole. In areas where probe holes cannot be made, metal rods are driven down to contact the main, and listening devices are applied to the rods.

The major advantages of the listening procedure for locating leaks are that no excavations are needed, distribution crews are not required to operate valves and hydrants, and the work can be performed easily by utility personnel.

Factors Affecting Leak Detection The following factors affect how well listening devices perform:

- Mains with rubber gasket joints often do not transmit sound much beyond the pipe section that has the leak.
- Copper transmits sound best, followed in order by steel, cast and ductile iron, plastic, asbestos cement, and concrete.
- Smaller pipes transmit sound well, but as the diameter increases, the quality of the sound diminishes.
- Tees, elbows, and other fittings often amplify sounds, making it difficult to locate a leak.
- Dry, sandy soils produce the best noise transmittance, whereas sounds are dissipated somewhat in loamy soils. Sounds are reduced even further in clay soils.
- Gravel roads and lawn areas are poor transmitters of sound compared with paved surfaces.
- The noise contributed by other buried utilities can also make it difficult to pinpoint a leak.

Water Audits

The purpose of a water audit is to select and implement a program to reduce distribution system water losses. The audit identifies the quantity of water lost and the cost. An added benefit is that records and control equipment (such as meters) are checked for accuracy. Before starting the audit, a worksheet to gather the information should be developed and the study period for the audit should be defined. Most utilities select a 12-month calendar year for this purpose. A final step before starting is to select the appropriate unit of measure. Many US utilities use gallons as the preferred measure.

A distribution system water audit consists of 11 main tasks (discussed in detail in AWWA Manual M36, *Water Audits and Loss Control Programs*):

1. Collect distribution system description information.
2. Measure water supplied to the distribution system.
   - Compile the volume of water from own sources.
   - Adjust figures for total supply. This step involves (a) verifying metered accuracy; (b) adjusting supply totals; (c) adjusting reservoir and tank storage; (d) making other adjustments; (e) totaling all adjustments; and (f) determining the adjusted volume of water from own sources.
   - Compile the volume of water imported from outside sources or purchased from other water utilities.
   - Calculate system input volume.
   - Compile the volume of water exported to outside water utilities or jurisdictions.
   - Calculate the volume of water supplied into the distribution system.
3. Quantify billed authorized consumption.
   - Compile the volume of billed authorized consumption (metered water). This step involves (a) maintaining customer accounts data; (b) maintaining customer meter and AMR data; (c) compiling metered consumption volumes for the water audit period; and (d) adjusting for lag time in meter readings.
   - Compile the volume of billed authorized consumption (unmetered water).

water audit
A procedure that combines flow measurements and listening surveys in an attempt to give a reasonably accurate accounting of all water entering and leaving a system.

4. Calculate nonrevenue water.
5. Quantify unbilled authorized consumption.
   - Compile the volume of unbilled authorized consumption (metered water).
   - Compile the volume of unbilled authorized consumption (unmetered water), which includes water for (a) firefighting and training; (b) flushing water mains, storm inlets, culverts, and sewers; (c) street cleaning; (d) landscaping irrigation in public areas; (e) decorative water facilities; (f) construction sites; (h) water quality and other testing; (i) water consumption at public buildings not included in the customer billing system; (j) other unmetered but verifiable uses; (k) sum of all components of unbilled authorized consumption that is unmetered.
6. Quantify water losses.
7. Quantify apparent losses.
   - Estimate customer meter inaccuracy. This step includes checking for proper installation, testing residential meters, and calculating total customer consumption meter error.
   - Estimate systematic data handling error, which includes systematic data transfer errors from customer meter reading, systematic data analysis errors, and policy and procedure shortcomings.
   - Estimate unauthorized consumption.
   - Calculate total apparent losses.
8. Quantify real losses.
9. Assign costs of apparent and real losses.
   - Determine cost impact of apparent loss components.
   - Determine cost impact of real loss components.
10. Calculate performance indicators.
    - Calculate the financial performance indicators.
    - Calculate the operational performance indicators, which includes apparent losses normalized, real losses normalized, infrastructure leakage index (ILI).
11. Compile the water balance.

After the audit, utilities analyze the value of the losses and corrective measures, evaluate potential corrective measures, update the audit, and update their master plan. An example summary from a water audit is shown in Table 8-2.

Emergency Leak Repairs

An emergency leak is usually a broken main or a severe service leak. An action plan to deal with main breaks should be established and coordinated with police, fire, and street department personnel. Trained personnel, records, maps, and repair parts should be available at all times in preparation for a major leak.

Preliminary Steps When a major leak is reported, it should be investigated at once to determine the severity of the problem. Utility personnel must determine if immediate protection of private and public property is needed, as well as whether there is any potential danger to pedestrians and traffic. Next, it should be determined, if possible, whether the repair is the responsibility of the utility or a property owner. If the leak is serious or in an area where it could quickly do extensive property damage, it must be repaired immediately.

Table 8-2 Typical water audit results

	Water Volume		
	ML/d	mgd	%
Water sold through domestic meters	9.4	2.5	50
Water sold through industrial meters	4.7	1.25	25
Underground leakage located	1.8	0.5	10
Underregistration of industrial meters	0.38	0.1	2
Unauthorized consumption	0.37	0.1	2
Total accounted-for water	16.65	4.45	89
Unmetered use (sewer flushing, firefighting, street washing)	0.18	0.05	1
Loss through unavoidable leakage at 200 gpd per mile (470 L/d per kilometer) of main	0.07	0.2	4
Unmetered use and underregistration of domestic meters	1.1	0.3	6
Total unaccounted-for water	1.35	0.55	11
Total water produced	18	5	100

The first consideration in dealing with the leak should be to get the water loss under control by complete or partial shutdown of the line. Some repairs can be accomplished while the pipeline is still under pressure. Maintaining a positive pressure in the line will ensure that backflow from cross-connections will not occur. Where damage to property is occurring or is likely to occur, water must usually be completely shut off.

If there is no immediate danger to life or property, all valves on that line except one should be closed to reduce the flow. Customers who will be without water should then be notified of approximately when the water will be shut off. They should also be advised to store some drinking water if they will need it during the outage. Special attention should be given to buildings with sprinkler systems, stores with water-cooled refrigeration units, industrial water users, and large users such as hospitals. These customers will have to take special steps to prepare for the water being off. If buildings are unoccupied, the owner, manager, or agent should be notified.

If the leak is at the bottom of a hill, customers at higher elevations should be notified to shut off the inlet valve at their meter to prevent siphoning out of hot-water tanks or softeners. If the occupants are not at home, the curb stops should be shut off.

In some instances, the utility may find it difficult to locate valves and mains because of incomplete records or snow cover. Having electronic locators on hand to locate metallic boxes and pipe can greatly facilitate a water main repair job (Figures 8-13 and 8-14).

The availability of information from a regular valve-locating and valve-exercising program will be helpful in an emergency situation. The valves that are operated and their operating conditions should be recorded. The section of the system that is out of service should be noted, and the locations of all hydrants affected should be reported to the fire department. After customers have been notified, the final valve can be shut down. Record all valve position changes so

Figure 8-13 Electronic pipe detector

Courtesy of Fisher Research Laboratory, Los Banos, California.

Figure 8-14 Valve and box locator

Courtesy of Fisher Research Laboratory, Los Banos, California.

that they can be returned to service following the repair. An up-to-date system map showing accurate valve locations and main sizes is valuable at this time.

Locating the Leak When water is flowing out of the ground, it may seem obvious where to look for a leak, but time and money can be wasted digging in the wrong spot. The leak is sometimes not directly below where the water comes to the surface. Occasionally, the water may show up a long way from the leak. A utility employee should pinpoint the leak using a listening device and probing the ground before starting to dig. Some water pressure will have to be left on or temporarily restored to the affected area during the leak-locating phase.

A large leak can undermine roads, walks, railways, or other utilities. In addition, basements may be flooded and slippery driving conditions may result from flooding or freezing water. It may be necessary to arrange for police assistance in controlling traffic and for street department assistance in salting or sanding roadways.

Before a leaking line is excavated for repair, other utilities must be contacted to determine where their underground lines are located. In some areas, all utilities cooperate in a single-call system that immediately alerts all utilities of the need for their assistance.

Excavation In addition to excavating equipment, the following items should be on hand in preparation for making the leak repair:

- A pump for dewatering the excavation
- A sturdy ladder for entering and leaving the excavation
- Shovels, wrenches, and other hand tools
- Planks and timbers for shoring trench walls
- Traffic control equipment, including barricades, flashers, and cones
- Pipe cutters or saws, as well as proper-size repair clamps, couplings, pipe, and sleeves
- An air compressor and hammer if pavement has to be cut or a clay spade for use in hand excavation

- A generator, lights, and flashlights if the work will be done at night
- Safety ropes or ladders if personnel must work where sudden flooding or a cave-in is possible
- A 5-gal (19-L) bucket with rope, to lower tools into the excavation

The excavation for repairing a leak should normally be parallel to the main or service. It should be located so that the pipe is to one side of the ditch. This will allow a worker to stand next to the pipe while making the repair. A sump hole should also be dug in a far corner of the excavation. The pump suction is placed in this hole to keep the bottom of the excavation as free of standing water as possible. The excavation should not be made too small in an attempt to save time. The extra time taken to make a neat hole that is large enough to work in will usually be more than made up for by the time saved in making the pipe repair quickly.

Safety procedures should not be ignored in the haste to handle the emergency. Accidental injuries are especially likely in unplanned situations. If soil conditions are poor or if the soil is unstable from being saturated by the leak, it may be necessary to install sheeting or shoring to protect workers from a trench cave-in.

Leak Repair If a break is severe, the damaged section may have to be cut away with a saw or pipe cutter and a new section installed. When a new section or appurtenance is inserted into a dewatered pipeline, the materials are disinfected with chlorine according to procedures described in AWWA Standard C651 (most recent edition). Most main breaks are straight across and can be repaired with a flexible clamp or coupling. Two types of repair devices are shown in Figure 8-15. The manufacturer's recommendations should be carefully followed when installing the repair device.

Several preliminary procedures should be observed to help ensure a quick and efficient repair. The diameter of the pipe should be checked to make certain the correct-size clamp is being installed. After the pipe is uncovered, the area where the clamp will be installed should be scraped and washed to remove as much dirt and corrosion as possible.

When applying the clamp, the workers should make sure no foreign material sticks to the gasket as the bottom half is brought around the pipe. No material should become lodged between the gasket and pipe as the bolts are tightened. A properly sized ratchet wrench should be selected in advance so that the sleeve bolts can be quickly tightened once the clamp is installed.

After the clamp is installed, some utilities apply a tar coating for rust prevention. The utility should then restore line pressure by partially opening one valve to test the repair for leaks before the excavation is backfilled. When the repair

backfill
(1) The operation of refilling an excavation, such as a trench, after the pipeline or other structure has been placed into the excavation. (2) The material used to fill the excavation in the process of backfilling.

Figure 8-15 Pipeline repair clamps

has been completed and tested, the main should be flushed to remove any air or dirt that may have entered while it was under repair. The entire line should be chlorinated to reduce the danger of contamination from soil, dirty water, and backsiphonage. The line is then flushed to remove high-chlorine water, and bacteria testing is conducted (follow requirements of AWWA Standard C651, most recent edition).

Restoration All unsuitable excavated material should be hauled away and replaced with suitable material for backfilling, with proper compaction. It is generally best to use sand, road base, crushed stone, or processed material for backfilling under pavement so that permanent street repairs can be made at once. Coordinate with the street department to facilitate repairs. The area should be checked to make sure no equipment or traffic hazards, such as ice or erosion, remain. All valves should be returned to operating position. Customers, the police, and the fire department should be notified that the repair has been completed. Customers may need to clear their service line by running cold water from an outside faucet.

Record Keeping A record of every water main break or leak repaired should be kept for future reference. The cause should be identified, such as shear break, pipe split, blowout, or joint leak. If any old pipe is removed during the repair, a piece should be kept and tagged with the location and date. This piece can be used for future reference on the condition of the pipe interior.

External Pipeline Corrosion

The soil surrounding buried pipe can cause external corrosion of some pipe materials if the pipe is not properly protected. Soil characteristics vary greatly from one area of the country to another or even from one part of a community to another. In general, asbestos–cement and concrete pipe will suffer harmful external corrosion only under aggressive soil conditions. Cast-iron and ductile-iron pipe do not need any additional protection in most soil conditions, but polyethylene wrap or other technologies may be required in aggressive soil. Steel pipe should be well coated and provided with cathodic protection under essentially all soil conditions.

The electrochemical corrosion process that takes place on the outside of metal pipes is very similar to internal corrosion. It involves a chemical reaction and a simultaneous flow of electrical current. Differences in the soil and the pipe surface create differences of electrical potential on the buried metal to form a corrosion cell. A direct electrical current flows from the anodic area to the cathodic area through the soil and returns to the anodic area through the metal.

Corrosion cells are generally created in metallic pipe by the surface impurities, nicks, and grains that are present. Oxygen differentials in the soil can also create corrosion cells. Severe corrosion of buried metal pipe can be serious enough to cause leakage, breaks, and a shortened service life for the pipe.

Corrosion frequently takes the form of pits in an otherwise relatively undisturbed pipe surface. These pits occur at the anodic areas and may eventually penetrate the pipe wall. Cathodic areas are generally protected in this process. This type of corrosion will occur on steel, cast-iron, and ductile-iron pipe.

Factors Affecting External Corrosion The degree of corrosion and the size of the area affected by corrosion depend on the quantity, size, and intensity of the corrosion cells. This intensity, in turn, depends on the corrosivity of the soil.

Some soil characteristics and conditions that are likely to increase the rate and amount of corrosion include the following:

- High moisture content
- Poor aeration
- Fine texture (e.g., clay or silty materials)
- Low electrical resistivity
- High organic material content (such as in swamps)
- High chloride and/or sulfate content, which increases electrical conductivity
- High acidity or high alkalinity, depending on the metal or alloy
- Presence of sulfide
- Presence of anaerobic bacteria

Most of these characteristics affect the electrical resistivity of a soil, which is a fair measure of a soil's corrosivity. The amount of metal removed from the pipe wall is proportional to the magnitude of the current flowing in the corrosion cell. The amount of current is, in turn, inversely proportional to the electrical resistivity of the soil or directly proportional to the soil conductivity. Therefore, a soil of low resistivity or high conductivity fosters corrosion currents and increases the amount of metal oxidized.

It has also been found that certain microorganisms in the soil can accelerate corrosion. Under anaerobic conditions (i.e., no oxygen present), iron bacteria or sulfate-reducing bacteria can accelerate corrosion.

Methods of Preventing External Corrosion Several solutions to the problem of corrosion have been tried. The solution that will be most effective in any situation will vary depending on the pipe material and the chemical and electrical conditions of the surrounding soil. Methods that have been found effective under certain conditions include the following:

- Specifying extra thickness for pipe walls
- Applying a protective coating
- Wrapping pipe in polyethylene plastic sleeves
- Installing cathodic protection on the pipe

Adding extra thickness to a pipe as a corrosion allowance is not an effective long-term remedy. Pipe corrosion is not uniform. Specifying a corrosion allowance for buried pipe only postpones the problem. Corrosion-prevention methods are much more cost effective.

Applying a hot bitumastic or coal-tar coating to the pipe exterior may not totally reduce soil corrosion. If there are any pinholes or breaks in the coating, bare metal at these locations will be anodic to the surrounding coated pipe. Corrosion will be concentrated in these areas. Coated pipe must be inspected and handled carefully to ensure that the coating is not damaged.

One way to control corrosion of cast-iron or ductile-iron pipe in corrosive soils is to encase the pipe in a polyethylene sleeve, as shown in Figure 8-16. When a tap is installed on pipe with a polyethylene sleeve, care must be taken to not expose the bare pipe around the completed tap. The method frequently used to easily prevent corrosion around the tap is to apply several wraps of heavy tape around the pipe in the area to be tapped before the tap is made. There is controversy on the effectiveness of poly-wrapping, as some corrosion engineers believe it makes things worse when water is present.

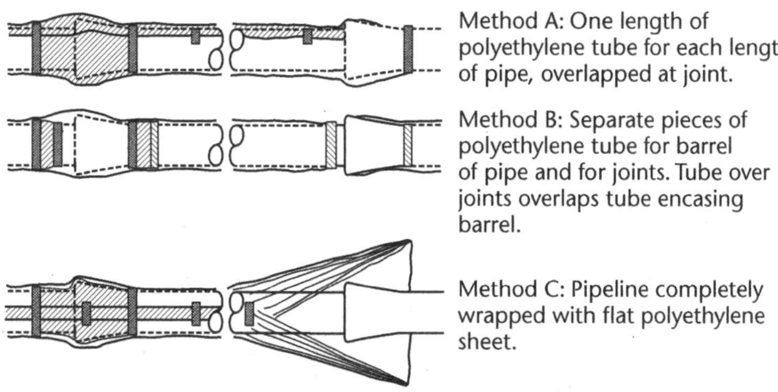

Figure 8-16 Polyethylene encasement of ductile-iron pipe

Courtesy of Farwest Corrosion Control Company.

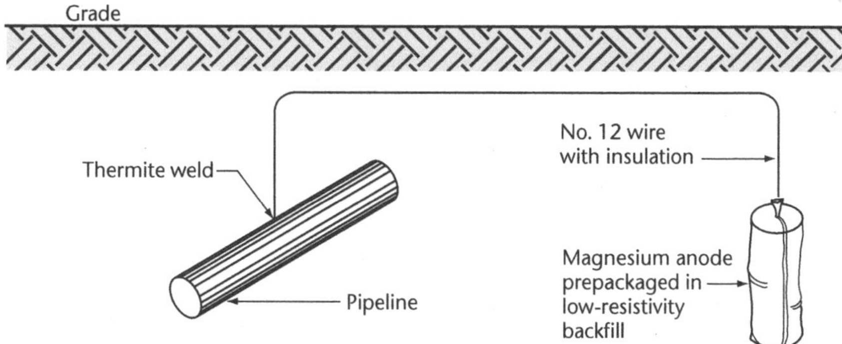

Figure 8-17 Cathodic protection using sacrificial anodes

Courtesy of Farwest Corrosion Control Company.

Cathodic protection stops corrosion by canceling out, or reversing, the corrosion currents that flow from the anodic area of a corrosion cell. There are two principal means of applying cathodic protection: sacrificial anodes and impressed-current systems.

Sacrificial anodes, also called galvanic anodes, consist of magnesium-alloy or zinc castings that are connected to the pipe through insulated lead wires, as shown in Figure 8-17. Magnesium and zinc are anodic to steel and iron, so a galvanic cell is formed that works opposite to the corrosion being prevented. As a result, the magnesium or zinc castings corrode (sacrifice their metal), and the pipe, acting as a cathode, is protected.

Impressed-current cathodic protection systems use an external source of direct current (DC) power that makes the structure to be protected cathodic with respect to some other metal in the ground. In most cases, rectifiers are used to convert alternating current (AC) power to DC. A bank of graphite or specially formulated cast-iron rods is connected into the circuit as the anode (corroding) member of the cell. The distribution system component is the protected cathode (Figure 8-18).

Cathodic protection is a very complicated and technical subject. A competent consultant should be involved early in any problem analysis or system revision.

cathodic protection
An electrical system for preventing corrosion to metals, particularly metallic pipe and tanks.

Bimetallic Corrosion

Another form of corrosion a water system operator may encounter is bimetallic corrosion. This type of corrosion occurs when two electrochemically dissimilar metals are directly connected to each other. The combination most often noticed

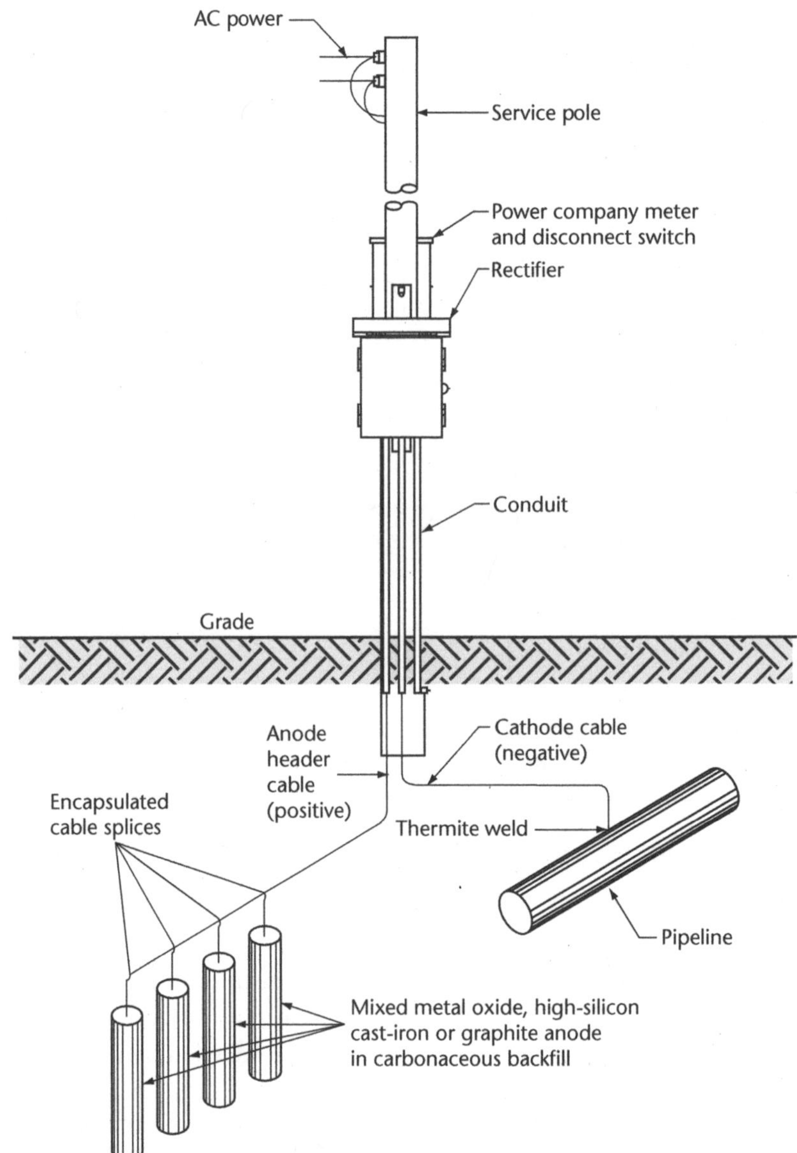

Figure 8-18 Cathodic protection using impressed current

Courtesy of Dr. J. R. Myers, JRM Assoc., Franklin, Ohio 45005.

galvanic cell
A corrosion condition created when two different metals are connected and immersed in an electrolyte, such as water.

galvanic corrosion
A form of localized corrosion caused by the connection of two different metals in an electrolyte, such as water.

in plumbing systems is the direct connection of a brass fitting to galvanized iron pipe. The two metals form a corrosion cell or, more exactly, a **galvanic cell** (Figure 8-19). **Galvanic corrosion** can result in loss of the anodic metal (in this case, the galvanized pipe) and protection of the cathodic metal (the brass fitting).

A practical galvanic series for common pipe and fitting materials is shown in Table 8-3. Each metal is anodic to, and may be corroded by, any metal below it. The greater the separation between any two metals in the series, the greater the potential for corrosion and (barring other factors) the more rapid the corrosion process.

Corrosion control methods for bimetallic connections are designed to break the electrical circuit of the galvanic cell. Dielectric barriers, or insulators, are the method most commonly used to stop the flow of corrosion current through the metal. In addition, coating a buried bimetallic connection reduces the amount of corrosion current passing through the soil.

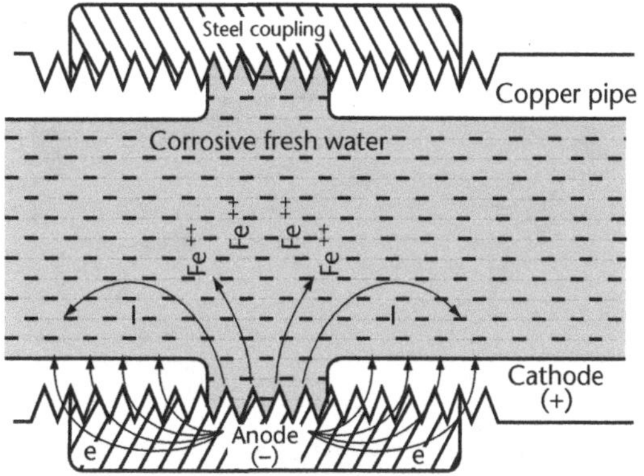

Figure 8-19 Small anode, large cathode

Table 8-3 Galvanic series for metals used in water systems

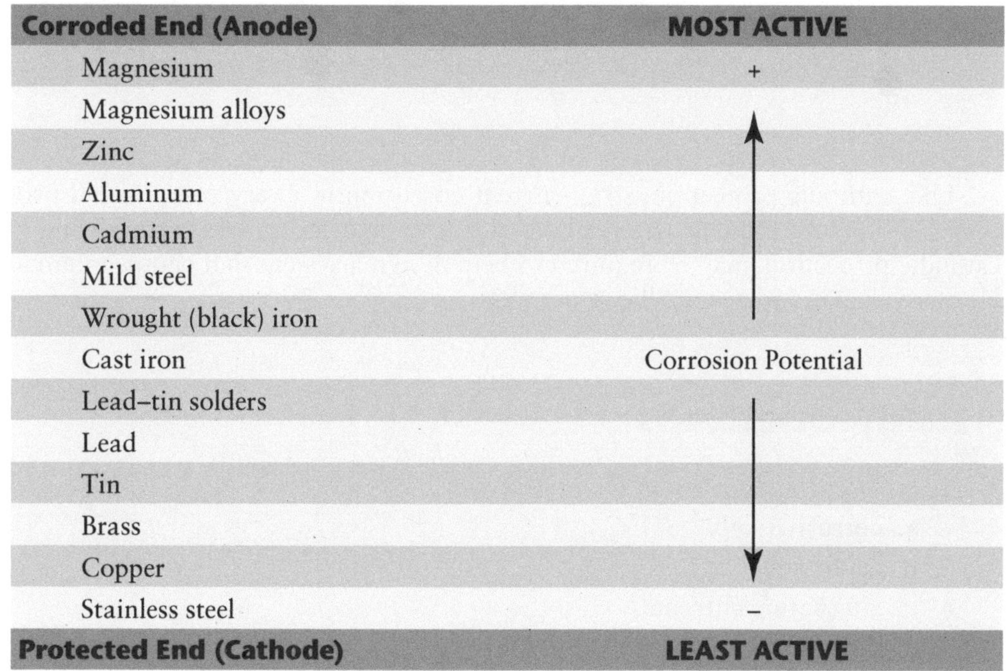

Corroded End (Anode)	MOST ACTIVE
Magnesium	+
Magnesium alloys	↑
Zinc	
Aluminum	
Cadmium	
Mild steel	
Wrought (black) iron	
Cast iron	Corrosion Potential
Lead–tin solders	
Lead	
Tin	
Brass	
Copper	↓
Stainless steel	–
Protected End (Cathode)	**LEAST ACTIVE**

Stray-Current Corrosion

Stray-current corrosion is caused by DC current that leaves its intended circuit, collects on a pipeline, and discharges into the soil. The classic problem that affected many old water systems involved currents generated by electric trolley cars and light rail systems, as shown in Figure 8-20. The same effect occurs as a result of DC current from any number of sources, including cathodic protection being applied to other utilities or structures, as well as subways.

Corrosion caused by stray-current discharge often appears as deep pits concentrated in a relatively small area on the pipe. If a main or service has this appearance, particularly at the crossing of a cathodically protected pipeline or transit facility, electrical tests should be conducted to determine if interference is the cause of the problem. Because many liquid-fuel and natural-gas pipelines are required to have cathodic protection, these lines become potential sources of stray current.

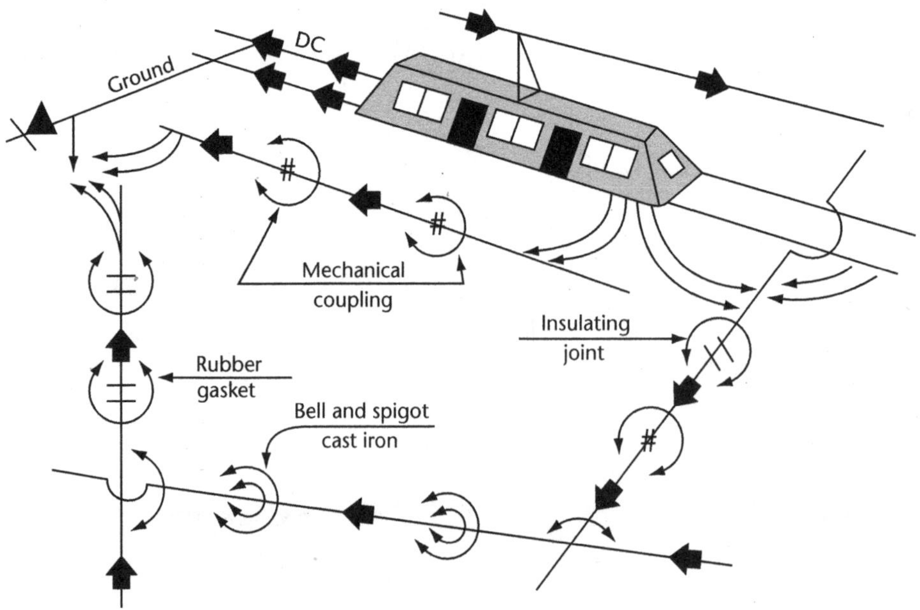

Figure 8-20 Stray current flow on underground pipelines

Like cathodic protection, stray-current corrosion is a very complicated process. Consultants and experts, available through engineering firms specializing in cathodic protection, may be required to help design a system that causes minimal interference with other metallic structures.

Study Questions

1. Stray-current corrosion is caused by
   a. corrosive soils.
   b. acidic soils.
   c. alternating currents.
   d. direct currents.
2. Sacrificial anodes are also called
   a. corrosion anodes.
   b. corrosion cells.
   c. galvanic anodes.
   d. decaying anodes.
3. Which of the following methods for preventing external corrosion is only a short-term remedy?
   a. Specifying extra thickness for pipe walls
   b. Wrapping pipe in polyethylene plastic sleeves
   c. Applying a coating of bitumastic or coal tar
   d. Installing cathodic protection on the pipe

4. In _____, air from a compressor is mixed with water and flushed through mains of less than 4 in. (100 mm) in diameter.
   a. slugging
   b. pigging
   c. backflowing
   d. air purging

5. After flushing mains, nozzle caps should be tightened to the point that
   a. they can easily be removed by hand, should an emergency arise.
   b. unauthorized persons cannot remove them by hand.
   c. a light cracking noise is heard.
   d. they cannot be removed without pneumatic tools.

6. What are the methods for leak detection?

7. What would be the best strategy when dealing with a distribution system area with positive coliform test results?

8. What is the term for a condition in which a pump, boiler, or other equipment produces a pressure greater than the water supply pressure?

9. What is the term for a corrosion condition created when two different metals are connected and immersed in an electrolyte, such as water?

10. List four changes that may occur in a distribution system to reduce (or increase) pressure and flow to an unacceptable level.

Chapter 9
Water Quality Testing

Microbiological Organisms

Indicator Organisms

The tests required to detect specific pathogens are still considered time intensive and expensive, so it is impractical for water systems to routinely test for specific pathogens. A more practical approach is to examine the water for indicator organisms specifically associated with contamination. An indicator organism essentially provides evidence of fecal contamination from humans or warm-blooded animals. The criteria for an ideal indicator organism are that it should

- always be present in contaminated water,
- always be absent when fecal contamination is not present,
- generally survive longer in water than do pathogens, and
- be easy to identify.

The coliform group of bacteria has been used for 100 years as an indicator of drinking water quality. These bacteria are generally not pathogenic, yet they may be present when pathogens are present.

Coliform bacteria are easily detected in the laboratory. As a rule, where coliforms are found in water, it is assumed that pathogens may also be present, making the water bacteriologically unsafe to drink. If coliform bacteria are absent, the water is assumed safe.

Many methods exist for determining the presence of coliform bacteria in a water sample, including the multiple-tube fermentation method, the presence–absence method, the MMO–MUG method, and the membrane filter method.

indicator organism
A microorganism whose presence indicates the presence of fecal contamination in water.

coliform bacteria
A group of bacteria predominantly inhabiting the intestines of humans or animals but also occasionally found elsewhere. Presence of the bacteria in water is used as an indication of fecal contamination (contamination by human or animal wastes).

Coliform Analysis

The detection of coliform bacteria in a water sample by any of the four analytical techniques is a warning of possible contamination. One positive test does not conclusively prove contamination, however, and additional tests must be conducted. Samples are often contaminated by improper sampling technique, improperly sterilized bottles, and laboratory error. Regulatory agencies recognize this fact, and drinking water regulations require further checking or repeat sampling after findings that show a positive test for coliform in a sample. Drinking water regulations and maximum contaminant levels (MCLs) for coliform bacteria are discussed in Chapter 1.

Sampling

Sterile containers must be used for all samples collected for bacteriological analysis. The same sampling procedures should be used for coliform analysis and heterotrophic plate count analysis. See Table 9-1.

Test Methods

The multiple-tube fermentation and presence–absence tests are designed on the principle that coliform bacteria produce gas from the fermentation of lactose within 24–48 hours when incubated at 35°C (95°F). Although the bacteria themselves cannot be seen, their presence is signified by the gas that is formed and trapped in an inverted vial in the fermentation tube.

Other testing methods that will be discussed include the MMO–MUG technique, the membrane filter method, and the heterotrophic plate count method.

Table 9-1 Total coliform sampling requirements, according to population served

Population Served	Minimum Number of Routine Samples per Month*	Population Served	Minimum Number of Routine Samples per Month*
15 to 1,000†	1‡	59,001 to 70,000	70
1,001 to 2,500	2	70,001 to 83,000	80
2,501 to 3,300	3	83,001 to 96,000	90
3,301 to 4,100	4	96,001 to 130,000	100
4,101 to 4,900	5	130,001 to 220,000	120
4,901 to 5,800	6	220,001 to 320,000	150
5,801 to 6,700	7	320,001 to 450,000	180
6,701 to 7,600	8	450,001 to 600,000	210
7,601 to 8,500	9	600,001 to 780,000	240
8,501 to 12,900	10	780,001 to 970,000	270
12,901 to 17,200	15	970,001 to 1,230,000	300
17,201 to 21,500	20	1,230,001 to 1,520,000	330
21,501 to 25,000	25	1,520,001 to 1,850,000	360
25,001 to 33,000	30	1,850,001 to 2,270,000	390
33,001 to 41,000	40	2,270,001 to 3,020,000	420
41,001 to 50,000	50	3,020,001 to 3,960,000	450
50,001 to 59,000	60	3,960,001 or more	480

Source: *Water Quality and Treatment* (1999).

*In lieu of the frequency specified in this table, a noncommunity water system using groundwater and serving 1,000 persons or fewer may monitor at a lesser frequency specified by the state until a sanitary survey is conducted and the state reviews the results. Thereafter, noncommunity water systems using groundwater and serving 1,000 persons or fewer must monitor in each calendar quarter during which the system provides water to the public, unless the state determines that some other frequency is more appropriate and notifies the system in writing. Noncommunity water systems using groundwater and serving 1,000 persons or fewer must monitor at least once per year.

†Includes public water systems that have at least 15 service connections but serve fewer than 25 persons.

‡For a community water system serving 25 to 1,000 persons, the state may reduce this sampling frequency if a sanitary survey conducted in the last 5 years indicates that the water system is supplied solely by a protected groundwater source and is free of sanitary defects. However, in no case may the state reduce the sampling frequency to less than once per quarter.

Multiple-Tube Fermentation Method The multiple-tube fermentation (MTF) test (Figure 9-1) progresses through three distinct steps:

1. Presumptive test
2. Confirmed test
3. Completed test

Presumptive Test The presumptive test is the first step of the analysis. Samples are normally pipetted into tubes containing a culture medium (lauryl tryptose broth [LTB]) with inverted filed vials containing the media in the tubes. The samples are incubated for 24 hours and then checked to see if gas has formed in the inner vial and cloudiness has developed in the broth. If neither is the case, they are incubated for 24 hours more and checked again.

If coliform bacteria are present in the water, the gas they produce will begin to form a bubble in each inverted vial within the 48-hour period; this is called a positive sample or reported as presence. If no gas forms, the sample is called negative or reported as absence. If gas is produced after either the 24-hour or the 48-hour incubation period, the sample must undergo the confirmed test.

Confirmed Test The confirmed test is more selective for coliform bacteria. This test increases the likelihood that positive results obtained in the presumptive test are caused by coliform bacteria and not other kinds of bacteria. Cultures from the positive samples in the presumptive test are transferred to brilliant green lactose bile (BGB) broth and incubated. If no gas has been produced after 48 hours of incubation, the test is negative and no coliform bacteria are present. If gas is produced, the test is positive, indicating the presence of coliform bacteria.

Bacteriological testing of most public water supplies stops after the confirmed test. This is the minimum testing that all samples must undergo when the MTF method is used. To check its procedures, the laboratory should conduct the completed test on at least 10 percent of the positive tubes from the confirmed test.

The sample may also be confirmed for the presence of fecal coliform or *E. coli* by using EC Medium or EC Medium with MUG, respectively. Follow the instructions in *Standard Methods for the Examination of Water and Wastewater* for the proper procedures.

Completed Test The completed test is used to definitely establish the presence of coliform bacteria for quality control purposes. A sample from the positive

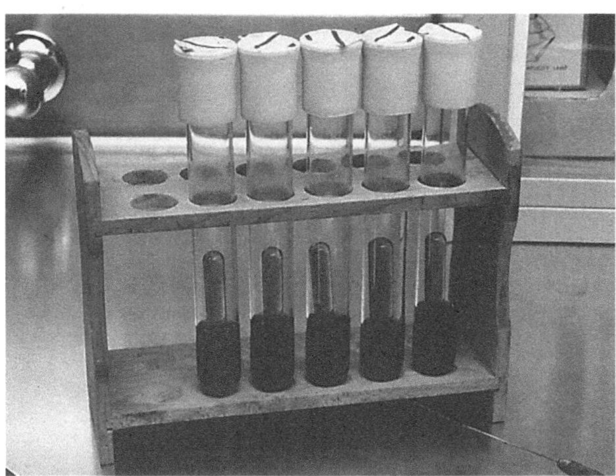

Figure 9-1 Typical multiple-tube fermentation setup
Source: Opflow.

multiple-tube fermentation (MTF) method
A laboratory method used for coliform testing that uses a nutrient broth placed in culture tubes. Gas production indicates the presence of coliform bacteria.

presumptive test
The first major step in the MTF test, which presumes (indicates) the presence of coliform bacteria on the basis of gas production in nutrient broth after incubation.

positive sample or presence
In reference to the MTF or membrane filter test, any sample that contained coliform bacteria.

negative sample or absence
When referring to the MTF or membrane filter test, any sample that does not contain coliform bacteria.

confirmed test
The second major step of the MTF test, which confirms that positive results from the presumptive test are due to coliform bacteria.

completed test
The third major step of the MTF test, which confirms that positive results from the presumptive test are due to coliform bacteria.

confirmed test is placed on an eosin methylene blue (EMB) agar plate and incubated. A coliform colony will form on each EMB plate.

A small portion of the coliform colony is transferred to a growth medium and incubated for 18–24 hours. A second portion is transferred to an LTB and incubated for 24–48 hours. The completed test is positive if (1) gas is produced in the LTB and (2) red-stained, non-spore-forming, rod-shaped bacteria are found. If no gas is produced in the LTB or if red-stained, chainlike cocci or blue-stained, rod-shaped bacteria are found on the agar, the test is negative. Figure 9-2 provides a summary of the MTF method. (The bacteria are not visible to the human eye. A microscope is needed to read the plates and determine the bacteria type.)

Presence–Absence Test The presence–absence (P–A) test is a simple modification of the MTF method. It is intended for use on routine samples collected from a distribution system or water treatment plant. A 100-mL portion of the

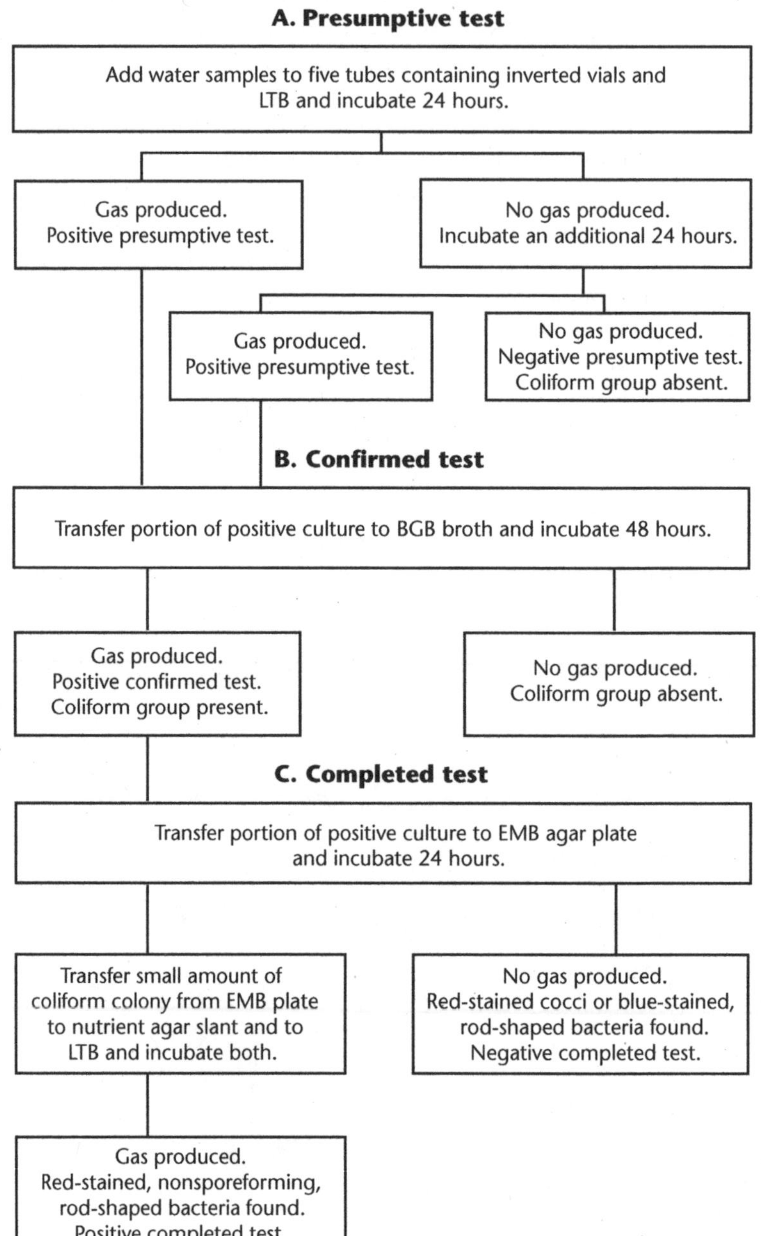

Figure 9-2 Summary of the multiple-tube fermentation method

presence–absence (P–A) test
An approved bacteriological procedure for the detection of total coliforms. The results are qualitative rather than quantitative.

sample is inoculated into a 250-mL milk dilution bottle containing special P–A media and a small inverted tube. The sample is then incubated at 35°C (95°F) for 24 and 48 hours.

The presence of total coliforms is indicated by the purple P–A medium turning yellow (indicating acid production) and by the formation of gas in the medium. All yellow and gas-producing samples from this presumptive stage must then be confirmed as described for the MTF-confirmed step using BGB tubes. Gas production indicates the presence of total coliforms and must be reported as a positive sample (presence) in the monthly report to the primacy agency.

Samples confirmed for total coliforms must also be analyzed for either fecal coliforms or *Escherichia coli*. A check or repeat sample must also be collected and analyzed. A check/repeat sample that is positive can result in an acute violation of the Total Coliform Rule (TCR) and must be reported to the primacy agency within 24 hours after results become known.

Fecal Coliform Procedure If the MTF or P–A method is being used, as the presumptive positive samples are being inoculated into the BGB broth, 0.1 mL of the presumptive broth is also transferred into an EC broth tube. (The actual name of the broth is EC, as it is a test for *E. coli.*) If the membrane filter method is used, bacterial growth is transferred into an EC tube. This tube is then incubated for 24 hours in a water bath at 44.5°C (112°F). The presence of gas in the tube confirms the presence of fecal coliforms.

E. coli Procedure The presence of *E. coli* can be determined using the MUG test discussed in the next section. A 0.l-mL portion of the presumptive media or a swab is used to transfer a sample from a membrane filter into an EC–MUG tube. A tube that fluoresces under a long-wave ultraviolet (UV) light is confirmation for *E. coli.*

MMO–MUG Technique The MMO–MUG technique was approved by the US Environmental Protection Agency (USEPA) shortly after promulgation of the TCR. MMO and MUG are acronyms for the constituents in the medium used in the tests. MMO represents minimal media with ONPG (ONPG stands for *ortho*-nitrophenyl-beta-D-galactopyranoside). *E. coli* produce a specific enzyme that reacts with ONPG to give a yellow color. MUG stands for 4-methylumbelliferyl-beta-D-glucuronide. Only *E. coli* produce an enzyme that reacts with MUG. Therefore, a medium containing MMO and MUG can be used to identify both total coliforms and *E. coli* in a single-sample inoculation.

Two procedures may be used. In the *ten-tube procedure*, ten tubes are purchased with the medium already in them. A 10-mL portion of sample is transferred into each tube and incubated at 35°C (95°F) for 24 hours. In the P–A procedure, the medium is purchased in vials. The medium is transferred into a bottle containing 100 mL of sample, is mixed, and is incubated as in the ten-tube procedure. If total coliforms are present in either procedure, the medium will turn yellow. If *E. coli* are present, the medium will also fluoresce blue under a UV light.

Membrane Filter Method The membrane filter method of coliform testing begins with the filtering of 100 mL of sample under a vacuum through a membrane filter. The filter is then placed in a sterile container or petri dish (Figure 9-3) and incubated in contact with a selective culture medium.

A coliform bacteria colony will develop at each point on the filter where a viable coliform bacterium was left during filtration. After a 24-hour incubation period, the number of colonies is counted (Figure 9-4).

MMO–MUG technique
An approved bacteriological procedure for detecting the presence or absence of total coliforms.

membrane filter method
A laboratory method used for coliform testing. The procedure uses an ultrathin filter with a uniform pore size smaller than bacteria—less than a micron. After water is forced through the filter, the filter is incubated in a special media that promotes the growth of coliform bacteria. Bacterial colonies with a green-gold sheen indicate the presence of coliform bacteria.

Figure 9-3 Placement of membrane on pad soaked with culture medium

Figure 9-4 Membrane filter after incubation with positive growth colonies

A typical coliform colony on M-Endo media is pink to dark red with a distinctive green metallic surface sheen. All organisms producing such colonies within 24 hours are considered presumptive coliforms. For confirmation, representative colonies are inoculated into LTB and BGB broth.

USEPA has revised the TCR, and the proposed new rule uses *E. coli* as the indicator for fecal contamination. Total coliform is used as an indicator that there may be a problem in the system requiring a system investigation.

Heterotrophic Plate Count Procedure The heterotrophic plate count (HPC) procedure is a way to estimate the population of bacteria in water. The test determines the total number of bacteria in a sample that grows under specific conditions in a selected medium.

Uses of the HPC Procedure No single food supply, incubation temperature, and moisture condition suits every type of bacterium being tested for, so a standardized procedure must be used to obtain consistent and comparable results. The procedure therefore generally permits only a fraction of the total population to be cultured. Often the number of HPC colonies is orders of magnitude lower than the total population present.

Plate-count tests are sensitive to changes in raw-water quality and are useful for judging the efficiency of various treatment processes in removing bacteria. For example, if a plate count is higher after filtration than before filtration, there may be bacterial growth on or in the filters. The problem would probably not show up during routine coliform analysis.

It is also common for water leaving a treatment plant to have a low bacterial population but for the population to have greatly increased by the time the water reaches the consumer. This occurrence is caused by bacterial aftergrowth (regrowth)—bacteria reproducing in the distribution system. Standard plate-count determinations may indicate whether this problem exists. Bacterial aftergrowth can generally result from water becoming stagnant in the dead ends in the system, inadequate chlorination, or recontamination of the water after chlorination.

heterotrophic plate count (HPC)
A laboratory procedure for estimating the total bacterial count in a water sample.

Performing the HPC Procedure The HPC is performed by placing diluted water samples on plate-count agar. The samples are incubated for 48–72 hours. Bacteria occur singly, in pairs, in chains, and in clusters. The bacteria colonies that grow on the agar are counted using colony-counting equipment. Detailed procedures are described in the latest edition of *Standard Methods*. These procedures must be closely followed to provide reliable data for water quality control measurements.

Properly treated water should have an HPC of less than 500 colonies per milliliter. Higher counts indicate an operational problem that should be investigated.

Alternate Methods Other methods for coliform and *E. coli* detection are being developed using a combination of enzymes, ß-glucuronidase, and ß-galactosidase in combination with ONPG and MUG. In tests that are positive for coliform, a yellow substance is produced that fluoresces at 366 nm UV light after an incubation period of 24 ± 2 hours at 35.0°C ± 0.5°C (~95°F). There are various methods of determination, from just adding the water sample to a test bottle containing the media and incubating, to pouring the water sample into a tray containing the detection media with volumetric cells for counting. The method has been approved for use by USEPA but may need to be confirmed with the drinking water primacy agency for your area.

Physical and Aggregate Properties of Water

Calcium Carbonate Stability

The principal scale-forming substance in water is calcium carbonate ($CaCO_3$). Water is considered stable when it will neither dissolve nor deposit calcium carbonate. This point is referred to as *calcium carbonate stability* or the *equilibrium point*. The reactions and behavior of calcium carbonate and calcium bicarbonate are therefore important in water supplies. The actual amount of calcium carbonate that will remain in solution in water depends on several characteristics of the water: calcium content, alkalinity, pH, temperature, and total dissolved solids.

Significance

Scale formation can cause serious problems in water distribution mains and household plumbing systems by restricting flow, plugging valves, and fouling water heaters and boilers. Corrosion can cause premature pipe or equipment failure. Public health and aesthetic problems can also result if water is corrosive, because pipe materials (e.g., lead, cadmium, and iron) will dissolve into the water.

Several methods can be used to determine the calcium carbonate stability of water. A popular method is the Langelier saturation index (LSI). The LSI is equal to the measured pH (of the water) minus the pH_s (saturation). The pH_s is the theoretical pH at which calcium carbonate will neither dissolve into nor precipitate from water. Water at the pH_s is considered stable. Therefore, if $pH - pH_s = 0$, the water is in equilibrium and will neither dissolve calcium carbonate nor deposit it on the pipes.

If $pH - pH_s > 0$ (positive value), the water is not in equilibrium and will tend to deposit calcium carbonate on mains and other piping surfaces. If $pH - pH_s < 0$ (negative value), the water is also not in equilibrium and will tend to dissolve the calcium carbonate it contacts; no coating will be deposited on the distribution pipes, and if the pipes are not protected, they may corrode.

The calcium carbonate stability of water is maintained in the distribution system by adjusting the LSI of the water to a slightly positive value so that a slight

calcium carbonate
Scale-forming substance in water.

equilibrium
A balanced condition in which the rate of formation and the rate of consumption of a constituent or constituents are equal.

deposit of calcium carbonate will be maintained on pipe walls. Adjustment is usually made by adding lime, soda ash, or caustic soda.

Indices other than the LSI use alkalinity as part of the equation or method to determine the stability of the water, especially its corrosiveness. One is the *marble test*, in which calcium carbonate (limestone) is dissolved in the water sample and the initial alkalinity is compared with the final alkalinity. The Ryzner index is used to perform a similar calculation to the LSI; it indicates the corrosiveness of the water compared to the pH_s.

Alkalinity is also important in determining how effectively the water reacts with coagulants for plant treatment. The alkalinity ions act as a "reservoir" of molecules that are available to react with coagulants or other chemicals such as disinfectant to reduce the pH change to the water that these chemical may affect. This effect is known as the "buffering capacity." Low-alkalinity waters have less buffering capacity and can experience wide pH fluctuations upon addition of coagulant, because many coagulant chemicals consume alkalinity. Coagulation and flocculation may be challenging in low-alkalinity waters and may require careful control to ensure effectiveness.

Sampling

If calcium carbonate stability maintenance is used for corrosion control, finished water at the treatment plant and in the distribution system should be evaluated routinely for calcium carbonate stability. Evaluation is particularly important when treatment plant unit processes or chemical doses are changed. If the LSI indicates unfavorable conditions, process adjustments should be made. It is very important to remember that the LSI is only an indicator of stability; it is not an exact measure of corrosivity or of calcium carbonate deposition. The LSI is developed from results of monitoring for alkalinity, pH, temperature, calcium content, and total dissolved solids (dissolved residue).

Methods of Determination

If the temperature, total dissolved solids, calcium content, and alkalinity of the water are known, the pH_s can be calculated. The following equation may be used:

$$pH_s = A + B - \log(Ca^{2+}) - \log(\text{alkalinity})$$

In the equation, A and B are constants, and calcium and alkalinity values are expressed in terms of milligrams per liter as calcium carbonate equivalents. Tables 9-2 and 9-3 are used to determine the values of the constants and logarithms.

The actual pH of the water is measured directly with a pH meter, and the LSI is calculated using the formula LSI = pH – pH_s. Drinking water samples are not typically diluted. A 1-mL sample of water is typically used for analysis.

Table 9-2 Constant *A* as a function of water temperature

Water Temperature, °C	*A*
0	2.60
4	2.50
8	2.40
12	2.30
16	2.20
20	2.10

Table 9-3 Constant *B* as function of total dissolved solids

Total Dissolved Solids, mg/L	*B*
0	9.70
100	9.77
200	9.83
400	9.86
800	9.89
1,000	9.90

Example

A sample of water has the following characteristics:

Ca^{+2}	=	300 mg/L as $CaCO_3$
Alkalinity	=	200 mg/L as $CaCO_3$
Temperature	=	16°C
Dissolved residue	=	600 mg/L
pH	=	8.7

Determine the saturation index (LSI):

$$pH_s = A + B - \log(Ca^{+2}) - \log(\text{alkalinity})$$

$$pH_s = 2.20 + 9.88 - 2.48 - 2.30$$

$$pH_s = 7.3$$

$$LSI = 8.7 - 7.3 = +1.4$$

An LSI of +1.4 indicates that this water is scale forming.

Inorganic Chemicals

Chlorine Residual and Demand

Chlorine is usually added to source water as the water enters the treatment plant (prechlorination) and again just before it leaves the plant (postchlorination). In the plant, chlorine is also often added at intermediate points during the treatment process. Postchlorination is primarily administered to provide an excess of chlorine for continued disinfection in the distribution system. Tests of chlorine levels in the plant and throughout the distribution system are necessary to determine that chlorine dosage levels are adequate and to monitor water quality.

Significance

Destruction of pathogenic organisms by chlorine is directly related to contact time and the concentration of the chlorine. High chlorine doses with short contact periods will provide essentially the same results as low doses with long contact periods. Chlorination also oxidizes substances such as iron, manganese, and organic compounds, making their removal from the water easier.

Successful chlorination requires that enough chlorine be added to complete the disinfection or oxidation process. However, chlorine must not be added in amounts that are wasteful, creating unnecessarily high operational costs and potentially promoting the formation of disinfection by-product compounds. Determining effective and efficient chlorine dosage levels is the responsibility of the plant operator.

Chlorine Residual

There are two types of chlorine residual: combined residual and free available residual. The process by which these are formed is illustrated in Figure 9-5.

The first amount of chlorine (e.g., 1 mg/L) that is added to raw water is used in oxidizing reducing compounds such as iron and manganese (from point 1 to point 2 in the figure). The chlorine oxidizes the iron and manganese and in the process is used up, meaning no residual forms and no disinfection occurs.

Breakpoint chlorination curve

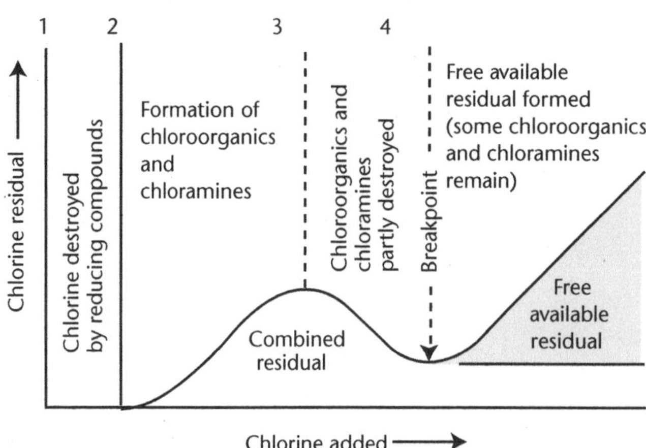

Figure 9-5 Formation of combined chlorine residual and free available chlorine residual

If the initial chlorine dosage used is higher (e.g., 2.5 mg/L), the reaction will go to point 3. Between points 2 and 3, the chlorine reacts with the organic substances and the ammonia in the water, forming chloroorganics and chloramines. These two products are called **combined chlorine residual**. This is a chlorine residual that—because it is combined with other chemicals, normally ammonia compounds, in the water—has lost some of its disinfecting strength. Compared with free chlorine, combined chlorine residual is a less effective disinfecting agent. If it is not properly controlled, it may cause tastes and odors characteristic of water in a swimming pool. Used in the appropriate applications and controlled correctly, monochloramine can be an effective distribution system disinfectant, frequently applied for the purposes of disinfection by-product control.

As the chlorine dosage is increased further (point 3 to point 4), the chloramines and some of the chloroorganics are destroyed. This process reduces the combined chlorine residual until, at point 4, the combined residual reaches its lowest point. Point 4 is called the **breakpoint**. At the breakpoint, the chlorine residual changes from combined to free available.

As the initial chlorine dosage is increased still further (beyond 4 mg/L in this example), free available chlorine residual is formed—free in the sense that it has not reacted with anything and available in the sense that it can and will react if necessary. In terms of disinfecting power, free available residual is 25 times more powerful than combined residual, and it will not produce the characteristic swimming-pool odor that combined residuals do. Because free available chlorine residual forms only after the breakpoint, the process is called **breakpoint chlorination**.

The free available chlorine residual at the consumer's tap should be at least 0.2 mg/L or at a level specified by the state. This level helps ensure that the water is free from harmful bacteria. However, higher levels may be necessary to control special problems, such as iron bacteria. If maintaining a free chlorine residual in a distribution system becomes difficult, several possible problems are indicated. Stagnant water in dead ends or storage tanks, biological growths, contamination of mains during main-break repair, and contamination caused by cross-connections all cause dissipation of free chlorine residuals. Further, a drop in chlorine residual in the distribution system may indicate inadequacies in the treatment process itself.

combined chlorine residual
The chlorine residual produced by the reaction of chlorine with substances in the water. Because the chlorine is "combined" it is not as effective a disinfectant as free chlorine residual. In water treatment, this usually refers to compounds formed by the combination of chlorine and ammonia.

breakpoint
The point at which the chlorine dosage has satisfied the chlorine demand.

breakpoint chlorination
The addition of chlorine to water until the chlorine demand has been satisfied and free chlorine residual is available for disinfection.

Chlorine Demand

Test results for chlorine residual can be combined with operating data regarding the amount of chlorine added at the plant to yield information on chlorine demand. Chlorine demand is a measurement of how much chlorine must be added to the water to achieve breakpoint chlorination or whatever free chlorine residual is desired.

The most significant reason for analyzing a water supply's chlorine demand is to determine the proper dosage. However, changes in chlorine demand can also indicate water quality changes. For example, if a water supply suddenly requires more chlorine to maintain a residual (that is, if the water exhibits a higher chlorine demand), then the chlorine is oxidizing some contaminants that previously were not present in the water supply.

When chlorine demand increases, two steps are necessary. First, the chlorine dose must be increased to meet the higher demand. Second, the reason for the increased demand should be investigated. A sudden increase in chlorine demand frequently occurs in surface water because of seasonal water quality changes. Chlorine demand in groundwater should not change substantially because the quality of groundwater is usually very stable.

Disinfection By-products

The disinfection by-products (DBPs) currently regulated are the trihalomethanes and the sum of five haloacetic acids (HAA5). Continuing studies and research have revealed that chlorine (and all other alternate disinfectants) reacts with organic compound precursors in the water to form many different kinds of organic compounds. In current thinking, many of these compounds are considered to be potentially toxic and are suspected of being carcinogenic. Haloacetic acids, halonitriles, haloaldehydes, and chlorophenols are just a few of the organic compounds associated with chlorine disinfection. Thus, chlorination has its good and its bad points; the water plant operator must know how to adequately disinfect the water without producing undesirable levels of DBPs.

Sampling

Chlorine residual sampling is performed at the treatment plant, at the consumer's faucet, and at strategic locations throughout the distribution system for both regulatory purposes and process control.

Treatment Plant Sampling In-plant sampling of chlorine residual determines whether sufficient chlorine has been added to the water before it leaves the treatment plant. This is the only way to be sure that finished water leaving the plant contains the desired chlorine residual. Obtaining representative samples is the most critical part of in-plant chlorine sampling.

In some instances, a sample must be collected at a point near the location of chlorine addition. In this case, analyses will probably show disproportionately high chlorine residual. To obtain data that approximate the actual chlorine residual in a particular basin, the sample should be held for a time period equal to the basin detention time or, in any case, at least 10 minutes. Under requirements of the Surface Water Treatment Rule, surface water systems serving populations larger than 3,300 are required to provide continuous chlorine residual monitoring where the water enters the distribution system.

Distribution System Sampling For systems that chlorinate, sampling for chlorine residual from the distribution system is done to determine whether consumers are receiving water that is of good quality. In other words, if there is no chlorine residual, the operator will have to determine the cause of the reduced residual and take corrective measures to restore the chlorine residual to the area to ensure the safety of the water supply. If the water samples are positive for bacteria and

chlorine demand
The quantity of chlorine consumed by reaction with substances in water.

chlorine residual is analyzed at the time of sample collection the information regarding the level of chlorine residual may aid in determining whether the contamination was in the water sample or whether the equipment used for the analysis was contaminated. A strong chlorine residual in the sample collected may indicate that the water was not contaminated.

If analysis is made in the field, only about 10 mL of sample is required. Current regulations require that chlorine residual samples taken for compliance samples must be analyzed immediately or at the most within 15 minutes of collection. If samples must be taken to a laboratory, a 100-mL sample should be collected. Analysis should be completed as soon as possible after collection. There is no recommended preservation for chlorine samples; chlorine is unstable in water and residual chlorine will continue to diminish with time, and so immediate field testing is preferred. Chlorine analyses performed after the 15-minute window are not considered acceptable results for compliance testing for regulatory purposes.

Agitation or aeration of the sample should be avoided because it can cause reduction of the sample's chlorine concentration. Chlorine will also be destroyed and subsequent analysis will be erroneously low if samples are exposed to sunlight. The same sample bottle should never be used for both chlorine residual and coliform analyses. Bottles used for coliform analysis contain a chemical (sodium thiosulfate) that neutralizes the chlorine residual.

Methods of Determination

The *N,N*-diethyl-*p*-phenylenediamine (DPD) test kit is the simplest and quickest way to test for residual chlorine. The test takes approximately 5 minutes to complete. The old orthotolidine method has been eliminated as an acceptable method and should not be used. It is not as accurate as the DPD method, particularly for measuring free chlorine residual.

Another technique, used primarily in laboratories because of its accuracy, is amperometric titration. The method is unaffected by sample color or turbidity, which can interfere with colorimetric determinations. However, performance of amperometric titration requires greater skill and care than does the DPD method. Because of the equipment and sample volumes required, the amperometric method is normally not used as a field test outside of the treatment plant.

The chlorine demand can be determined by treating a series of water samples with known but varying chlorine dosages. After an appropriate contact time, the chlorine residual of each sample is determined. This procedure indicates which dosage satisfied the demand and provided the desired residual.

Organic Contaminants

Measurement of Organic Compounds

No single analytical method is capable of measuring all of the organic substances in a water sample. However, available analytical methods can be grouped into two categories, general and specific.

General Analytical Methods

Threshold odor tests, flavor profiles, and color determinations have been used in the water utility industry for many years to obtain general measures of the levels of natural organic compounds in water. Two other methods used occasionally in monitoring water quality are UV light absorbance and fluorescence. These tests are used in some plants for control of organic compound removal processes because the measurement can be made quickly and easily.

Another test commonly used to determine the overall content of organic compounds in water is the measurement of total organic carbon (TOC). The typical concentration of TOC in water sources ranges from less than 0.5 mg/L to more than 10 mg/L. Highly colored water may have a TOC concentration of more than 30 mg/L.

Total organic halogen (TOX) is another measurement that is being used increasingly because it is specific to halogenated organic compounds. The presence of TOX in a sample is an indication of the presence of either synthetic organic compounds or DBPs. Only fairly sophisticated laboratories are currently capable of carrying out the procedures for TOC and TOX determinations.

Specific Analytical Methods

The list of organic compounds that have been identified in drinking water samples has grown from approximately 200 in 1975 to thousands today, and it is constantly lengthening. In many cases, of course, a compound may have been identified only in isolated samples or at extremely low concentrations. However, the growth in the list is primarily because of steadily increasing improvements in analytical methods.

Gas chromatography (Figure 9-6) requires very specialized equipment, detailed procedures, and trained operators. The three fundamental steps are as follows:

1. Extraction and concentration of the organic compounds in the sample
2. Separation of the extracted organic compounds in a gas chromatograph
3. Detection of individual compounds

Extraction and Concentration Organic substances are first extracted from a water sample. One method uses an organic solvent such as methylene chloride or pentane. This process is called liquid–liquid extraction. Another method strips

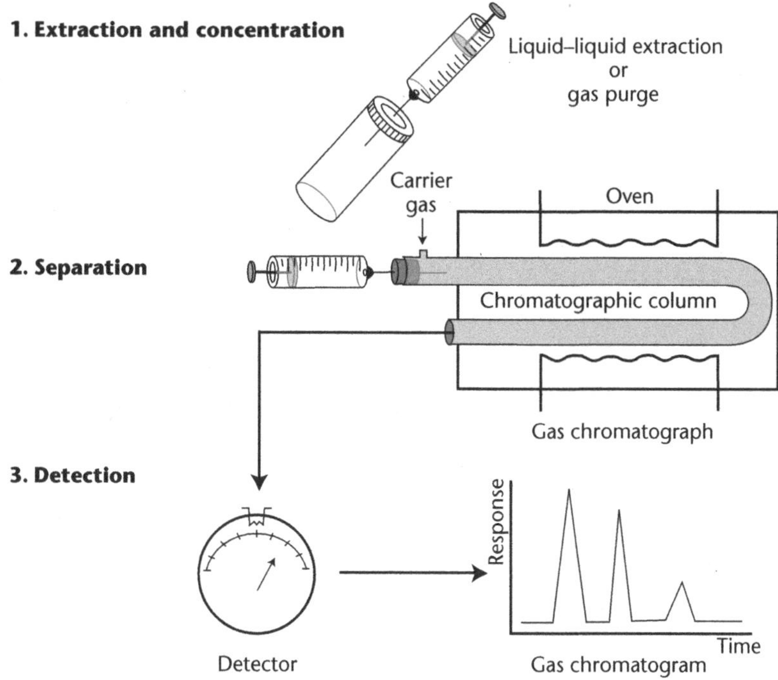

Figure 9-6 Steps in gas chromatography

total organic carbon (TOC)
The results of a general analysis performed on a water sample to determine the total organic content of the water.

the organic compounds out of the sample using an inert gas such as nitrogen or helium. This process is called purge-and-trap analysis.

Separation The complex solution must then be separated into its individual organic components. This process is carried out with a gas chromatograph or a high-performance liquid chromatograph. Chromatographs have a column of long, thin tubing through which individual organic compounds are driven off the sample as the temperature is elevated. Thus, these processes may be viewed as sophisticated distillation or separation functions.

Detection

As the chromatograph separates the organic compounds by the temperature at which they are vaporized, they travel to a detector. Several types of detectors are available, each with certain advantages and disadvantages. The types in general use include flame ionization, electron capture, electrolytic conductivity, photoionization, and mass spectrometry. An organic compound is identified by comparing the signal the detector obtains (shown graphically in a gas chromatograph [GC] generated by the detector) with known standards for the compound. This process is generally aided by a computer connected to the equipment.

Figure 9-7 shows a chart produced by a GC, showing the presence of trihalomethanes in a water sample.

Sampling for Organic Compounds

The location of sampling points in a water distribution system is very important; certain points should definitely not be used. The following are among typical locations to be avoided:

- Public restrooms should not be used as sampling locations because the deodorizer commonly used in restrooms contains an organic chemical that may be in the air in sufficient concentration to contaminate the water sample.

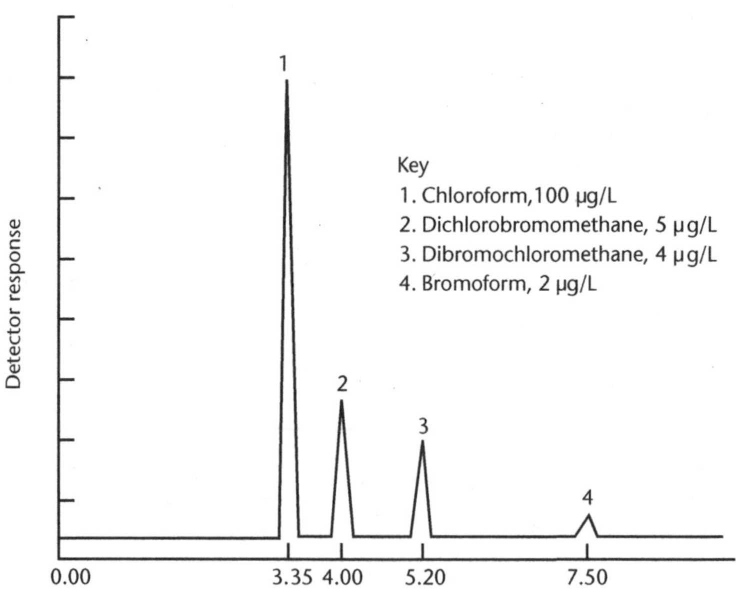

Figure 9-7 Sample readout from a gas chromatograph

gas chromatograph (GC)
A technique used to measure the concentration of organic compounds in water.

- Gasoline service stations should be avoided because of the prevalence of petroleum products that could be in the air or that could have gotten on the sampling faucet.
- Any location where there are unusual odors, such as a freshly painted room, or where there is a smell from cleaning materials, should be avoided.
- A location where a pump or piping has recently been installed or repaired, especially if organic solvents have been used for cleaning and degreasing, should be avoided because of the possibility that organic solvents may have been used in the plumbing.
- Other unsuitable locations are those where solvents may be present in the atmosphere such as paint or hardware stores, barber and beauty shops (hair spray, etc.), and dry-cleaning establishments.

Ultraclean glass vials having lids with polytetrafluoroethylene (PTFE; trade name Teflon) liners are used for collecting volatile organic compound samples. The samples must be collected so that there is zero headspace in the vial; in other words, there must be no bubble of air in the vial after it is filled. If any air remains in the vial, a portion of the more volatile organic compounds will come out of solution and into the air space, which will cause inaccurate analysis of the water sample. Trip blanks and field blanks should also be incorporated into the normal sampling practices.

Each sample container must be completely labeled. A general rule is that the description of the sampling site must be complete enough so that a person unfamiliar with the initial sampling could return and collect a repeat sample from the same location if necessary. Refer to *Standard Methods for the Examination of Water and Wastewater* for more information.

Radiological Contaminants

One of the more significant public health concerns regarding drinking water is the relatively high level of natural radioactivity found in some water sources. Most radioactivity in water occurs naturally, but there is also a threat of radionuclide contamination from various industrial and medical processes.

The harmful effects to a living organism of consuming water containing radioactivity are caused by the energy absorbed by the cells and tissues of the organism. This absorbed energy (or dose) produces chemical decomposition of the molecules present in the living cells. Each of the forms of radiation reacts somewhat differently within the human body.

Radioactive Materials

A radioactive atom (Figure 9-8) emits alpha particles, beta particles, and gamma rays.

Alpha Particles (Radiation)

Alpha particles are the most prevalent naturally occurring radionuclide present in drinking water and are therefore of the greatest concern. Alpha particles are the heaviest particles.

Alpha radiation is not true electromagnetic radiation like light and X-rays. It consists of particles of matter. Alpha particles are doubly charged ions of helium. Although they are propelled from the nucleus of atoms at approximately 10 percent of the speed of light, they do not travel much more than 10 cm in air at room

radioactivity
Behavior of a material that has an unstable atomic nucleus, which spontaneously decays or disintegrates, producing radiation.

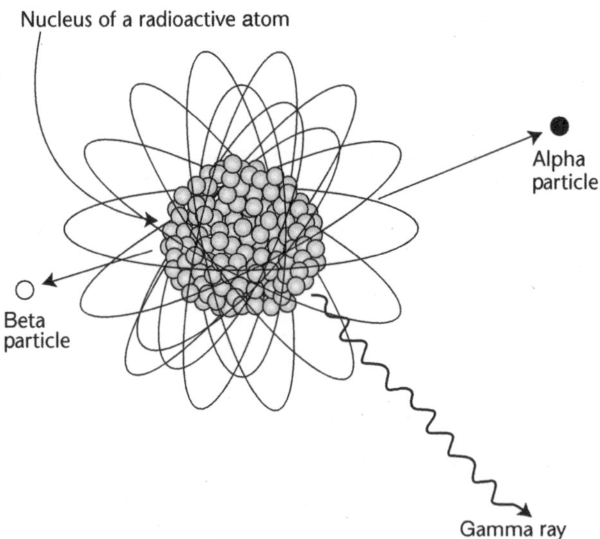

Figure 9-8 Emissions from the nucleus of a radioactive atom

temperature. They are stopped by an ordinary sheet of paper. The alpha particles emitted by a particular element are all released at the same velocity. The velocity varies, however, from element to element. Alpha particles have extremely high ionizing action within their range.

Beta Radiation

Beta radiation consists of negatively charged particles—*electrons*—that move at speeds ranging from 30 to 99 percent of the speed of light. The penetrating power of beta radiation depends on its speed. It can travel several hundred feet in air and can be stopped by aluminum a few millimeters thick. The ionizing power of beta radiation is much less than that of alpha radiation.

Gamma Radiation

Gamma radiation is true electromagnetic radiation, which travels at the speed of light. It is similar to X-ray radiation but has a shorter wavelength and therefore greater penetrating power, which increases as the wavelength decreases. Proper shielding from gamma radiation requires a barrier of lead that is several centimeters thick or concrete several feet thick. The unit of gamma radiation is the *photon*.

Unit of Radioactivity

The measurement of radioactivity disintegration is expressed in *curies*. Formerly, one unit of radioactivity was considered to be the number of disintegrations occurring per second in one gram of pure radium. Because the constants for radium are subject to revision from time to time, the International Radium Standard Commission has recommended the use of a fixed value, 3.7×10^{10} disintegrations per second, as the *standard curie* (Ci).

The curie is used mainly to define quantities of radioactive materials. A curie of an alpha emitter is the quantity that releases 3.7×10^{10} alpha particles per second. A curie of a beta emitter is the quantity of material that releases 3.7×10^{10} beta particles per second, and a curie of a gamma emitter is the quantity of material that releases 3.7×10^{10} photons per second. The curie represents such a large number of disintegrations per second that the *millicurie* (mCi), *microcurie* (µCi), and *picocurie* (pCi)—corresponding to 10^{-3}, 10^{-6}, and 10^{-9} Ci, respectively—are more commonly used.

The *roentgen* is a unit of gamma or X-ray radiation intensity. It is of value in the study of the biological effects that result from ionization induced within cells by radiation. The roentgen is defined as the amount of gamma or X-ray radiation that will produce in one cubic centimeter of dry air, at 0°C (32°F) and 760 mm pressure, one electrostatic unit (esu) of electricity. This is equivalent to 1.61×10^{12} ions pairs per gram of air and corresponds to the absorption of 83.8 ergs of energy.

The roentgen is a unit of the total quantity of ionization produced by gamma radiation or X-rays. Dosage rates for these radiations are expressed in terms of roentgens per unit of time.

With the advent of atomic energy involving exposure to neutrons, protons, and alpha and beta particles—which also have effects on living tissue—it has become necessary to have other means of expressing ionization produced in cells. Three methods of expression have been used.

The **roentgen equivalent physical (rep)** is defined as the quantity of radiation (other than X-rays or other generated radiation) that produces in one gram of human tissue ionization equivalent to the quantity produced in air by one roentgen of radiation or X-rays (equivalent to 83.8 ergs of energy). The rep has been replaced largely by the term **radiation absorption dose (rad)**, which has wider application.

The rad is a unit of radiation corresponding to an energy absorption of 100 ergs per gram of any medium. It can be applied to any type and energy of radiation that leads to the production of ionization. Studies of the radiation of biological materials have shown that the roentgen is approximately equivalent to 100 ergs/g of tissue; it can be equivalent to 90–150 ergs/g of tissue depending on the energy of the X-ray radiation and type of tissue. The rad, therefore, is more closely related to the roentgen than is the rep, in terms of radiation effects on living tissues, and is the term biologists prefer.

The rad represents such a tremendous radiation dosage, in terms of permissible amounts for human beings, that another unit has been developed specifically for humans. The term **roentgen equivalent man (rem)** is used. The rem unit corresponds to the amount of radiation that will produce an energy dissipation in the human body that is biologically equivalent to one roentgen of radiation of X-rays, or approximately 100 ergs/g. The recommended maximum permissible dose for radiation workers is 5 rem/year; for nonradiation workers it is 0.5 rem/year.

roentgen equivalent physical (rep)
The quantity of radiation (other than X-rays or other generated radiation) that produces in one gram of human tissue ionization equivalent to the quantity produced in air by one roentgen of radiation or X-rays (equivalent to 83.8 ergs of energy).

radiation absorption dose (rad)
A measure of the dose absorbed by the body from radiation (100 ergs of energy in 1 g of tissue).

roentgen equivalent man (rem)
A quantification of radiation in terms of its dose effect on the human body; the number of rads times a quality factor.

Radioactive Contaminants in Water

Humans receive a radiation dose of about 200 millirems (mrem) or 0.2 rem from all sources each year, and USEPA estimates that on average as much as 3 percent of this dose comes from drinking water. Local conditions can, of course, greatly alter this proportion.

The following are some of the radioactive substances currently listed for testing as potential drinking water contaminants:

- Radium
- Uranium
- Radon
- Artificial radionuclides

Radium

Radium is the most common radionuclide of concern in drinking water. Naturally occurring radium leaches into groundwater from rock formations, so it is present in water sources in those parts of the country where there is radium-bearing rock. It may also be found in surface water as a result of runoff from mining and industrial operations where radium is present in the soil. The three *isotopes* (variations) of radium of concern in drinking water are radium 226, which emits principally alpha particles; radium 228, which emits beta particles and alpha particles from its daughter decay products; and radium 224, which has a very short half-life of about 3.6 days compared with radium 226 and radium 228, whose half-lives are measured in years. Currently, federal regulations ignore radium 224, but some states require monitoring for this isotope even though the sampling, shipping, and testing requirements make it difficult to obtain meaningful results.

Uranium

Naturally occurring uranium is found in some groundwater supplies as a result of leaching from uranium-bearing sandstone, shale, and other rock. Uranium may also occasionally be present in surface water, carried there in runoff from areas with mining operations. Uranium may be present in a variety of complex ionic forms, depending on the pH of the water.

Radon

Radon is a naturally occurring radioactive gas that cannot be seen, smelled, or tasted. Radon comes from the natural breakdown (radioactive decay) of uranium. It is the direct radioactive-decay daughter of radium 226. The highest concentrations of radon are found in soil and rock containing uranium. Significant concentrations, from a health standpoint, may be found in groundwater from any type of geologic formation, including unconsolidated formations.

Outdoors, radon emitted from the soil is diluted to such low concentrations that it is not of concern. However, when it is liberated inside a confined space, such as a home or office building, radon can accumulate to relatively high levels, and inhalation of the gas is considered a health danger. Most cases of excessive levels of radon in buildings are caused by the gas seeping through cracks in concrete floors and walls. In areas where high levels of radon in the soil are a problem, foundation ventilation should be installed to reduce the concentration of radon entering buildings.

The problem from a public water supply standpoint is that, if radon is present in the water, a significant amount of the gas will be liberated into a building as water is used. Showers, washing machines, and dishwashers are particularly efficient in transferring radon gas into the air. The radon released from the water adds to the radon that seeps into a building from the soil, adding to the health risk.

Artificial Radionuclides

Significant levels of artificial radionuclides have been recorded in surface waters as a result of atmospheric fallout following nuclear testing, leaks, and disasters. Otherwise, surface water generally contains little or no radioactivity. Potential sources of serious water contamination are accidental discharges from facilities using radioactive materials, such as power stations, industrial plants, waste-disposal sites, or medical facilities. State and federal nuclear regulatory agencies monitor all uses of radioactive materials to prevent such discharges. If an accidental discharge of artificial radionuclides takes place, the elements most likely to be present are strontium 90 and tritium.

Adverse Health Effects of Radioactivity

The effects of excessive levels of radioactivity on the human body include developmental problems, nonhereditary birth defects, genetic effects that might be inherited by future generations, and various types of cancer. All radionuclides are considered to be carcinogens.

Radium is chemically similar to calcium, so about 90 percent of naturally occurring radium that is ingested goes to the bones. Consequently, the primary risk from radium ingestion is bone cancer. Although uranium has not definitively been proven to be carcinogenic, its potential to accumulate in the bones, much as radium 228 does, has prompted USEPA, as a policy matter, to consider uranium a carcinogen and set the MCLG at zero. The principal adverse effect of uranium is toxicity to human kidneys.

Inhaled radon is considered to be a cause of lung cancer. Radon is also thought to have some noncarcinogenic effects on internal body organs when ingested. Although the proportion of radon added to a building by the water supply is usually relatively small in comparison with the amount that seeps into the building from the soil, the issue of radon in drinking water is still significant because of the many people being exposed. USEPA estimates that between one and five million homes in the United States may have significantly high levels of radon contamination and that between 5,000 and 20,000 lung cancer deaths a year may be attributed to all sources of radon. USEPA has not set, as of this writing, an MCL for radon in drinking water.

Radionuclide Monitoring Requirements

The level of restrictions that should be placed on radioactivity in drinking water has been the subject of extensive research and much debate. Some experts feel that the requirements should be much more restrictive, and others believe the danger is not serious and the requirements should be relaxed.

Another factor that has contributed to the dilemma of regulation is the high cost of radionuclide analyses. Although some progress has been made in the form of improved equipment and automated operation, analyses still require expensive equipment and trained staff to operate it. The cost is kept as low as possible by requiring an initial scan to determine if significant radioactivity is present. Only if the level is higher than a specified point, which is normally the detection limits for gross alpha or gross beta emitters but can be varied by the drinking water primacy agency, would it be necessary to progress to further analyses.

Customer Inquiries and Complaint Investigation

Responding positively to customers' inquiries and concerns about water quality can be very beneficial for water utilities in many ways. The process of responding gives operators an additional tool in tracking water quality through the distribution system and determining what customer concerns are. A customer complaint may be an early indication of a quality problem. In addition, customers' confidence in the quality of the water may be strengthened by professionally conducted investigations into their concerns.

The previous chapters have established the basis for gathering the data and improving the operator's knowledge of water quality and the effectiveness of the treatment process. These data are also useful in providing customers with the information they need to assure themselves that the water they are receiving is safe. In addition, the data provide the information needed to produce the annual consumer confidence report as required by regulation (see Chapter 1).

A telephone discussion with—and, if needed, a visit from—a well-informed utility employee can certainly improve the utility's public image for customers. Such discussions or visits are also opportunities to educate customers regarding the utility's operations and water quality.

Water treatment professionals should keep themselves informed not only about changes in regulations and government press releases, but also about any media articles relating to possible water quality, such as main breaks, boil-water orders, and outbreaks or potential outbreaks of diseases in the area or nation. Staying up to date will be better prepare employees to handle customers' questions and concerns.

General Principles

Inquiries regarding water quality should be handled promptly and courteously. Many of these are health- or aesthetics-related such as the quantity of fluoride, hardness, trihalomethane levels, or other parameters. Others may be as simple as whether the water source is surface water or groundwater. These may be driven by a request from a medical person for treatment or by news articles or discussions with relatives or friends.

When a complaint is received, an investigation should be undertaken according to the philosophy that the customer would not be calling if there were not a problem. On average, a water utility receives complaints from approximately less than 1 percent of its customers; thus the system may have a larger problem than the number of callers reflects. The problem might be real or only perceived; regardless, the customer has a problem and would like it resolved. The caller may be angry, frustrated, embarrassed, or uncomfortable about calling, so the receiver of the call should allow the customer to explain the reason for calling before starting to ask questions. Always obtain the customer's name and address at the beginning of the discussion, which offers reassurance that you are interested and are focused on the area of concern. This is a good approach for customers who are in a highly emotional state, because it gives them the opportunity to vent their feelings so they can discuss the subject calmly.

The first order of business is to define the problem. The customer may not know how to explain exactly what is bothering him or her or exactly what questions to ask. The receiver should repeat to the customer what he or she has heard: "You said, Mrs. Smith, that the water coming from your water faucets is brown and has an odor?" or "You indicate that your family is ill from drinking the water?" With the problem established and agreed on, further questions may be asked to gain details.

Complaint or Inquiry Form

It is helpful for the receiver to have an electronic or paper form to fill out while taking the complaint or inquiry. The form should have three parts:

1. The receiving information, including name, address, phone number, date, and time and type of complaint
2. The investigation results, including lab results
3. A description of the final disposition, including the customer's satisfaction with the investigation

Investigation

Although it is often difficult, the investigator should approach the problem with an open mind, having no preconceived notions about any part of the investigation. The customer should be asked again to explain the problem, and the investigation should be limited to that problem only.

In most investigations, water samples should be collected at a cold-water tap before any customer treatment, either to confirm the problem or to prove that the water being delivered to the premises matches the general condition of the water in the distribution system. Temperature and chlorine residual should be tested on-site, and a general chemical sample (hardness, pH, alkalinity, and the like) and a bacteriological sample should be collected for analysis in the lab.

If the solution to the problem is obvious, the customer should be informed immediately. If the solution is the customer's responsibility, the investigator should advise the customer about ways and means to implement the solution. If the solution is the utility's responsibility, the investigator should advise the customer how the utility will deal with the problem if possible. If the problem is a perceived one—that is, not really a problem—the investigator must communicate this information tactfully to the customer.

Final Disposition

Regardless of the details of the investigation, the investigator should carry it through to a resolution. Customers should be notified of any laboratory results, kept advised of the investigation's progress, and contacted on its conclusion to ascertain their satisfaction. The final result should be a satisfied customer. The completed complaint form should be kept on file for future reference.

Specific Complaints

A vast majority of complaints fall into one or more of the following categories:

- Objectionable taste and/or odor
- Objectionable appearance of the water
- Stained laundry and plumbing fixtures
- Illness alleged to be caused by the drinking water

Taste and Odor

Surveys have shown that taste-and-odor complaints are the type received most frequently by most water utilities, especially utilities treating surface waters. Sources of compounds that cause taste-and-odor problems may be natural or may be caused by pollution. Natural compounds result from plant growth including algae or animal activities in the watershed or source water. Most such natural compounds produce fishy, earthy, or manure-type tastes and odors. Industrial and agricultural/residential discharges into water sources generally produce chemical or medicinal tastes and odors.

Human perceptions of tastes and odors are highly variable. How individuals define what they taste and smell depends on many factors, such as the person's age, health, previous experiences, level of sensitivity, and other senses as they interact with taste and smell. For example, people will not say an odor "smells like a potato bin" if they have not smelled a potato bin. And it is not unusual for people to think they detect an odor in cloudy or colored water because of what they see. All these factors make the investigation of taste-and-odor complaints very tricky.

Receiving Information Once it has been established that taste and odor are the subjects of a complaint, customers should be asked to describe what they taste or smell and what the source appears to be—for example, hot or cold faucets, kitchen sink, bathroom sink, or bathtub. It should also be determined at this point if there are customer-owned water treatment devices installed in the line and how long the problem has been evident. The information obtained during this conversation should be passed on to the person performing any subsequent investigation.

Investigating The investigator should try smelling and/or tasting the water from the same faucets in which the customer has noticed the condition. To protect the investigator from a possible contaminant in the home that could affect his or her health, the recommendation is to smell first and taste (using a clean laboratory container) only if the odor is acceptable. If no taste or odor is detected in the cold water, water from the hot-water system can be tested. Many taste- and odor-causing compounds are volatile and can be tasted or smelled more readily from hot water. It may also be possible that the taste or odor exists only in the hot water, in which case the problem can be immediately located in the residence's hot-water system.

If a customer detects an odor when it is already known that the water system is experiencing a taste-and-odor episode, allow the customer to describe the taste or odor to ensure it is from the known condition. In such a case, the source of the problem can be explained to the customer over the phone. If a customer's detection of an odor appears to be an isolated case, further investigation is required. If the investigation reveals no odor, more investigation may still be necessary to convince the customer that the problem is only a perceived one. In any case, samples should be collected for study at the plant or laboratory.

In conducting the investigation, the investigator should attempt to imagine the many potential sources of taste and odor. Following is a list of some of the more probable causes and situations.

- A general taste-and-odor incident is occurring in the source water, and the caller is the first customer who has complained.
- The customer's water service is connected to a dead end or to a low-flow main in the distribution system, and stagnant water is being drawn into the residence's water service.
- A cross-connection has drawn some foreign substance into the water system.
- Water system maintenance in the vicinity has stirred up stagnant water or sediment in the mains.
- Waste plumbing or the trap under the sink or bathtub is the actual cause of the odor problem.
- The taste or odor is originating in the hot-water system in the residence.
- If a home water conditioner (water softener, carbon filter, or the like) is being used, it could be causing the problem.
- Customers who are in poor health or are elderly may be more inclined to imagine a problem.
- Customers may actually be tasting or smelling something that is not in the water supply (e.g., medication they are taking).
- The customer may be noticing the effect of some recent plumbing work in the building that resulted from a change in piping or from the cleaning solution the plumber or heating contractor used.

- There may be a cross-connection from a sink, garden hose, or flush valve in a toilet, or contractor in the area may be using a system hydrant and allowing foreign material into the customer's water system.
- Electrical problems may also be causing a problem, either a bad ground in the customer's building or something outside the home such as a power utility grounding condition (after a lightning storm) or from stray current from electrical rail lines, trolleys, or passenger trains in the area.
- After the terrorist attacks of September 11, 2001, utilities must also consider the possibility of deliberate contamination, which makes the investigation a priority.

Disposition of the Complaint After the problem has been identified, appropriate action must be taken. If the problem is within the jurisdiction of the utility, corrective measures must begin as soon as possible. If the problem is isolated within a residence, the investigator must work with the customers by advising them of the steps they can take to eliminate the problem. In all cases, customers must be kept advised of all the steps being taken to solve the problem, including the results of any laboratory testing. After the problem has been solved, customers should be contacted to verify their satisfaction with the situation.

Physical Appearance

Customers generally expect clear, odorless water to be available from their taps at all times. When the water deviates from this norm, they become concerned and report their concern to the water utility. The physical appearance of water can be affected adversely by such things as excess air in the water, sediment from disturbed water lines, rust, particulate matter, bugs, or worms. The latter two have been noted in systems providing drinking water from unfiltered surface water supplies (some large cities, because of their watershed protection programs, are allowed to operate unfiltered surface water systems) or, in the past, from areas where the finished-water reservoirs are open to the atmosphere. Currently all finished-water reservoirs must be covered. A purveyor that has been granted an exemption must provide filtration and disinfection before water from the uncovered reservoir reaches the public.

Receiving Information The receiver of the complaint call needs to obtain an exact description of the offending appearance of the water and when the customer first noticed it. On some calls, the receiver may be able to diagnose the problem and offer assistance over the phone, particularly if the cause is already known from a previous investigation in the same area or if there is a general condition being experienced at the time.

Investigating The investigator must observe the appearance of the water to develop information from which to draw conclusions and offer solutions. In some cases, it may be necessary to analyze material in the water chemically or under a microscope. When the material and the source have been identified, the investigator can take steps to correct the problem.

One rather common complaint received by some water systems is that the water is "cloudy" when all it contains is entrained air. Sometimes the callers are new customers who had not seen entrained air in the water where they lived before, but complaints may also come from old customers who may just be noticing the phenomenon for the first time. The problem occurs at certain times of the year, especially in cold-weather months, when the water becomes saturated with air.

Because a cold pressurized liquid holds more gas than a warm liquid when the water warms in the customer's building, the "extra" air is released, similarly to how it is released from a water heater. When a glass of water is filled, bubbles are released, making the water look cloudy when it is fresh from the tap. The cloudiness quickly clears, though, starting almost immediately at the bottom of the glass and moving upward; the water is completely clear in a minute or so. This cloudiness can also be the result of a defective faucet aerator or a throttled valve causing a restriction in the pipeline and a drop in pressure that releases air. Another source of cloudy water can be bad check valves in air compressors or compressed gas cylinders tied into water lines such as you would find in medical facilities or food vending locations.

Disposition of the Complaint Most complaints of dirty or discolored water due to dissolved or suspended matter in the mains can be cleared up by flushing the distribution system and the customer's plumbing. Regardless, once the problem and source have been identified, the investigator must follow through to a conclusion. Again, the utility must take action if the problem falls under its jurisdiction, and the investigator should suggest solutions to the customer if the problem is isolated in the residence.

Staining of Laundry and Plumbing Fixtures

Staining of laundry and plumbing fixtures can occur when the water contains iron, manganese, or copper in solution. It is relatively common for there to be some dissolved iron and manganese.

When a groundwater system pumps directly from wells to the distribution system, the water is generally clear as it comes from the customer's tap. However, after the water is exposed to air in a bathtub, toilet, or washing machine, iron oxidizes to red-brown ferric hydroxide precipitate. In some situations, iron is partially or completely oxidized in the water mains, and customers get discolored water either continuously or sporadically.

The iron precipitate causes laundered white clothes to have an off-white color, and brown stains build up on porcelain fixtures. A particularly exasperating problem for customers is that, as they repeatedly scour the porcelain fixtures to remove the discoloration, they slowly break down the porcelain's surface glaze, exposing the more porous ceramic below; these areas then become discolored even faster and are harder to clean. In this case there is generally no recourse but to replace the fixtures.

Another common complaint from customers when the iron content is high is that coffee and tea turn so dark they look like ink. This darkness is caused by a reaction of the iron with the tannic acid in the beverages.

Manganese is often present with iron in groundwater and may cause similar staining problems, except that a dark-brown to black staining precipitate is formed.

Copper staining is usually most objectionable when it creates blue-green stains on plumbing fixtures. Copper staining is caused by aggressive water that dissolves copper from the customer's piping system. Copper release from the customer's piping can also be caused by stray electrical currents, such as a bad ground or other situation (e.g., an appliance or water fountain with electrical cooling); under these conditions the exterior of the copper piping may develop a black coating.

Receiving Information The receiver of the complaint call needs to obtain a description of the problem and the customer's location. If the cause is already known from previous complaints, it may be possible to give the customer advice over the phone about removing the stains or preventing future stains.

Investigating If the complaint is new for the system or for a particular area of the system, the investigator should visit the customer and observe the problem. In some cases, staining can occur as a result of a local problem such as a dead-end main, and special corrective action may be possible.

Disposition of the Complaint If the problem is only a local condition, it may be possible to correct it by flushing mains in the area. If the problem recurs regularly in the area, it may be necessary to set up a regular schedule for flushing the mains.

If the problem is found to be general throughout the system, the utility should take steps to provide treatment to prevent staining from occurring.

Illness Caused by Water

Contaminated drinking water can cause illness, and customers generally have been made aware of this fact through information from educational institutions, the media, or their doctor. Some customers call the water utility after visiting a doctor who says that one source of their illness could be drinking water. Depending on the type of illness, the doctor or medical facility is obligated to report the findings to health authorities up to and including the Centers for Disease Control and Prevention. The utility may then be dealing with the local, county, state, or federal agency assigned to investigate the alleged incident.

Receiving Information Calls concerning waterborne illness may be some of the most difficult to handle. In many cases, the customer is unsure of terminology and does not know what questions to ask to initiate the investigation. The receiver of the complaint must be very sensitive in attempting to gain information. Very seldom, with the exceptions of *Giardia* and *Cryptosporidium*, will the infectious agent be known.

The chances of illness being caused by contamination of a well-run public water system are quite remote, but it does happen, so the customer's complaint cannot be immediately discounted. The receiver needs to determine the symptoms of the illness, the number of people in the household who are affected, and whether the illness has been diagnosed by a physician. The receiver must be careful not to sound as if he or she has medical knowledge when responding to or asking questions.

Investigating Generally, in customer-initiated calls, the customers are seeking to learn whether drinking water is a potential source of their illness. The investigator's job is to provide enough information that customers can reach their own conclusions regarding water quality.

Even if the person has an illness that is known as a waterborne disease, such as giardiasis or cryptosporidiosis, the illness could have been contracted through a source other than the water system. The vast majority of cases are actually contracted through person-to-person contact, although an occasional case of giardiasis can be traced to a person's contact with untreated water during a camping or fishing trip.

Nevertheless, a sample for bacteriological analysis should be drawn from a cold-water tap along with a sample for general chemical analysis to set the customer's mind at ease. A chlorine residual test should be conducted in the presence of the customer and the results explained. The investigator should tell the customer that the bacteriological analysis will determine the presence or absence of coliform bacteria and that these coliform bacteria are an indicator for the possible presence of pathogenic microorganisms in the supply. The investigator should further explain that the chemical analysis will compare the customer's tap water with the water being sent from the distribution system. This comparison will allow the investigator to determine whether a problem is occurring in that part of the system or in the person's residence, possibly due to a cross-connection. An innocent-looking water line penetrating an outside wall may be connected to an undocumented alternate water source, or contamination could occur from an unknown customer filter, or mixed plumbing—such as a sprinkler system or treatment device.

The investigator should also explain to the customer what precautions are taken to protect the water supply and how the water is treated by the utility. The customer should be encouraged to consult with a physician if that has not been done.

If investigations indicate the possibility that a waterborne illness is occurring, it is prudent and necessary that the utility notify the primacy agency and possibly also the state and/or local health departments and request their assistance.

Disposition of the Complaint The results of the bacteriological and chemical analyses should be relayed to the customer as soon as possible and a discussion held as to the customer's perception of the investigation. The complaint form should be filed for future reference.

WATCH THE VIDEO
Taste & Odor (www.awwa.org/wsovideoclips)

Study Questions

1. The radiation term corresponding to an energy absorption of 100 ergs per gram of any medium is called a(n)
   a. rad.
   b. rem.
   c. curie.
   d. erg.

2. In reference to the multiple-tube fermentation or membrane filter test, any sample that does not contain coliform bacteria is said to be
   a. negative.
   b. neutral.
   c. undefined.
   d. positive.

3. Which of the following is a technique used to measure the concentration of organic compounds in water?
   a. MMO–MUG technique
   b. Gas chromatograph
   c. Presence–absence (P–A) test
   d. Membrane filter method

4. When receiving a customer complaint about water quality, which of the following is a good approach for customers who are in a highly emotional state?
   a. Let the customer know you are already aware of the problem and your service has initiated action.
   b. Tell the customer to call back after calming down.
   c. Ask the customer to collect a sample and send it to your facility with expedited delivery.
   d. Obtain the customer's name and address at the beginning of the discussion.
5. What is the waiting period to receive a coliform test result?
6. How can the chlorine residual decrease when more chlorine is added to a water sample?
7. Why can't samples be preserved for delayed determination of pH?
8. What is the third major step of the multiple-tube fermentation method?
9. How long does the *N,N*-diethyl-*p*-phenylenediamine (DPD) test kit take to complete?

Chapter 10

Information Management and System Mapping

Automated Mapping/Facility Management/Geographic Information Systems

An automated mapping/facility management/geographic information system (AM/FM/GIS) collects, stores, manipulates, and analyzes water system components for which geographic location is an important characteristic. Information about public and private properties; road networks; topography; distribution and transmission lines; and facilities such as tanks, pump stations, pipes, wells, and treatment plants are all tied to geographic location.

In an AM/FM/GIS, geographic data are represented as points, lines, and areas on a computerized map. The computerized map has many layers, each layer corresponding to related data, such as topography, streets, water distribution, or electric power system. The concept is similar to a map with many transparent overlays but much more precise and detailed. In addition to maps, the data may be presented in the form of tables or even as text descriptions. The AM/FM/GIS may have hundreds of factors associated with each feature or location.

Basic Elements of an AM/FM/GIS

The basic elements of an AM/FM/GIS to handle geographic information include data entry (collection, management, and storage), data manipulation (modeling and analysis), and data output (retrieval, management, display, and storage). Table 10-1 lists the attributes of some data management systems.

Data entry requires converting data from an existing form into one that can be used by an AM/FM/GIS. As with any engineering system, the results are only as accurate as the initial data. The initial data entry is a complicated and labor-intensive procedure and thus is expensive.

The data manipulation function includes modeling and analysis. An AM/FM/GIS can automate certain activities within the organization. It will also change the way the organization works. For example, an AM/FM/GIS can quickly generate more alternatives for a project, helping with decision making. An AM/FM/GIS is not merely a high-tech way to produce system maps.

The output functions, determined by the user's requirements, include the retrieval, management, display, and storage functions. Output will be in the form of printed tables and reports, plotted maps and drawings, or digital files.

automated mapping/facility management/geographic information system (AM/FM/GIS)
A computerized system for collecting, storing, and analyzing water system components for which geographic location is an important characteristic.

Table 10-1 Attributes of data management systems

GIS	AM/FM	OSS/DSS/BSS
Macroview	Microview	Date intensive
Small scale; 1:2,000 typical	Large scale; 1:50, 1:100, 1:200 typical	Tabular
Spatial analysis	Network analysis	Aspatial analysis
Spatial topology required	Network topology required	No topology
Not time sensitive	Time sensitive	Mixed time sensitivity
More static	Dynamic	Historical
Not mission critical	Often mission critical	Often mission critical
Few users	Many users	Many users

AM/FM = automated mapping/facility management, BSS = business support system, DSS = decision support system, GIS = geographic information system, OSS = operational support systems.

System Software

The software used depends on the size of the database, whether data are to be shared among multiple users, and what functions will be performed. The size of the database dictates the level of sophistication required for the software to locate the requested information quickly. The need to share common data requires the software to distribute data to multiple users accurately and quickly. The types of functions to be performed determine the database management architecture, as well as the data structure and the functionality required. Figure 10-1 shows a complex software configuration.

Water Utility Applications

The objective of an AM/FM/GIS is to provide a fully integrated database to support management, operation, and maintenance functions at all levels. The applications can range from the daily transactions of preparing work orders to repairing meters to providing analyses of alternative design plans for new water main locations. Table 10-2 enumerates the typical water utility data sets, and Figure 10-2 shows the layered database categories that are commonly used. Table 10-3 lists some of the applications the AM/FM/GIS would provide for a water utility.

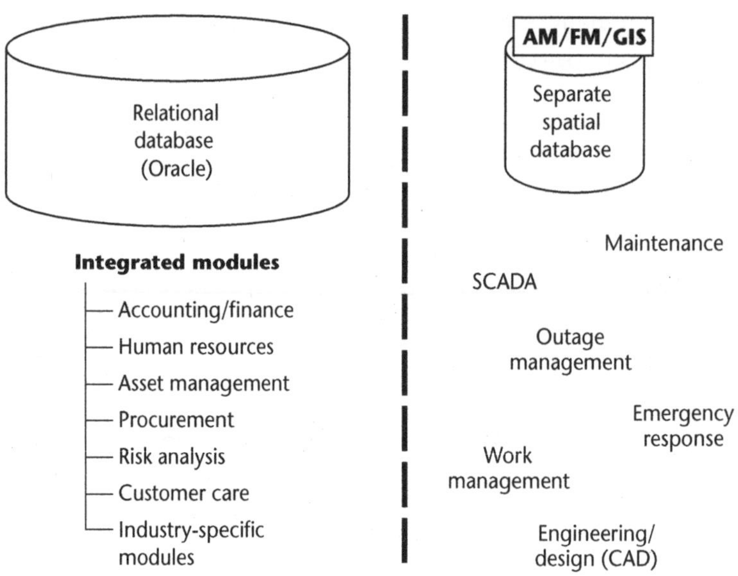

Figure 10-1 Complex software configuration for a water utility

Table 10-2 Typical water utility data sets

Data Category	Example Map Layers
Base data	Control information Planimetric features Hydrology features
Facilities and distribution systems data	Water piping Water valves and utility holes Service areas Water plant facilities (wells, pump stations, tanks, treatment plants) Other utilities
Land records data	Property boundaries Easements Right-of-ways
Natural resources data	Groundwater data Drainage data Soils data Flood plain boundaries Topographical features Vegetation information
Transportation network data	Road center lines Pavement locations Road intersections Bridges

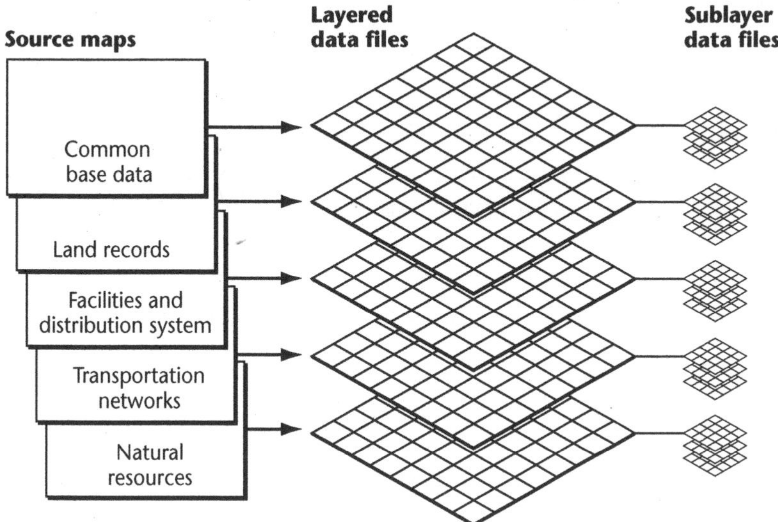

Figure 10-2 Typical layered database categories

Table 10-3 Typical water utility applications

Type of Application	Example
Facility management	Update, display, and analyze facility data; support the planning, design, operations, and maintenance to water facilities
Emergency response	Provide display of emergency vehicle routing; analyze frequency and location of emergency event
Area mapping and reporting	Analyze and display maps; produce maps and reports
Facility inspection	Schedule and track inspection; perform safety violations inspections; log and process complaints
Permitting and licensing	Process and track information
Land-use and environmental planning	Display and analyze land-use and environmental data such as water quality information by source
Facility siting	Select optimal locations for new utility facilities
Code enforcement	Schedule and track code violations; display and analyze enforcement data
Customer service	Analyze complaints, billing errors, accounts receivable, water usage by location and demographics

Water Resource Management

The AM/FM/GIS can create distribution system maps showing water quality data (e.g., bacteria counts, heavy metal concentrations) overlaid and related to the location of water production (e.g., wells, treatment plants). This graphical representation is used by the operator to control pumping and treatment facilities based on location within the community and water network, so that high-quality water is available to the customer. Planning personnel also use water quality trends, again overlaid onto the graphical model of the network, to analyze any degradation of water sources in relation to other variables such as consumption, season, or weather.

Customer Services

Information on billing data and meter numbers can be graphically represented and located on distribution system maps. Analyzing such things as accounts receivable information, consumption, and complaints with respect to street address, sector in the community, and location in the water network provides important information about the effectiveness of service, system reliability, and customer satisfaction.

Operations

An AM/FM/GIS connected with a supervisory control and data acquisition (SCADA) system is an effective tool for utility management. For example, suppose a control center operator receives a trouble call from a customer. The AM/FM/GIS links the customer's address or meter number to the collection or distribution system piping schematic (Figure 10-3). The pump station status is available on a SCADA display. An informed decision, based on current data, is made by the central operator, and crews are dispatched quickly with adequate information to resolve the problem.

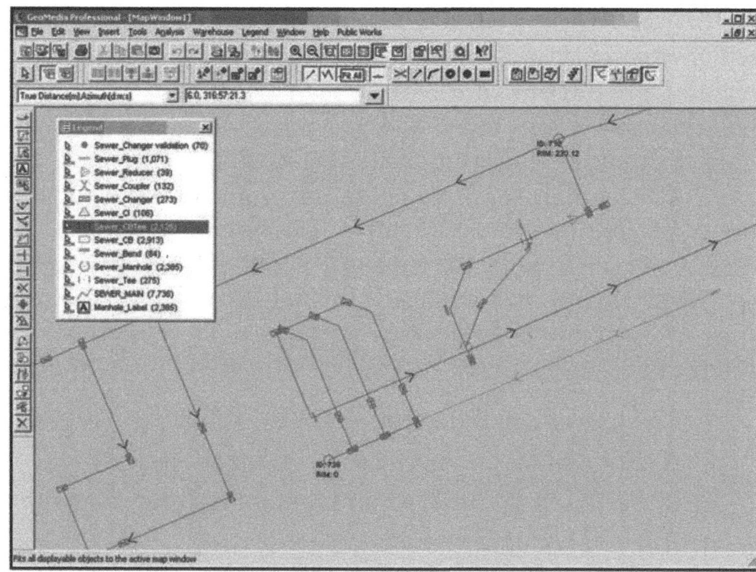

Figure 10-3 Customer records linked with mapping information

Courtesy of Integraph Mapping & Geospatial Solutions.

Water Demand Forecasting

Utilities can graphically represent the utility system on an AM/FM/GIS for short-term demand forecasting. For example, water demand is highly dependent on rainfall, because customers tend to decrease or stop irrigation when it rains. Plotting the relationships between rainfall measurements, map location, and demand provides useful historical information when a weather-sensitive, short-term demand forecast needs to be made. The utility can then use the demand forecast to help establish a pumping schedule based on economics or reliability.

Water System Modeling

An accurate and precise AM/FM/GIS distribution system network database can be accessed automatically by the planning model, resulting in more precise modeling. Overlaying accurate data with data from the SCADA system also provides a powerful tool for water system analysis.

Maintenance Management

In addition to the basic records discussed previously, certain other types of records are important. Work orders are used to communicate between the office and the field, and technical information on installed or available equipment must be available for maintenance and design.

Work Orders

Distribution system records are only as accurate as the information recorded. Therefore, every system must have an established procedure by which field crews complete and submit accurate information on the work that they do. A good working relationship between field and office personnel will help ensure a reliable record system.

Many water distribution systems use a **work order** to instruct field crews on specific work to be done. Work orders are also used for other purposes, such as ordering materials, organizing field crews, and obtaining easements.

work order
A form used to communicate field information back to the main office. Also used to order materials, organize field crews, and obtain easements.

Blanket work orders, which cover general repairs that are expected to fall within a specific time frame or dollar amount (as opposed to costly or unplanned repairs), are often used for work such as grounds keeping, changing light bulbs, painting, safety meetings, and special projects. They track maintenance hours used for indirect and nonspecific maintenance activities. This enables managers to effectively plan work, schedule staff, and track estimated-versus-actual time usage, permitting more accurate scheduling in the future. Blanket work requests are usually initiated by maintenance supervision or management.

When the job is completed, the field crew completes the form to show that the job was completed. A work order usually contains information such as the following:

- The location and description of the job to be performed
- A listing of all the materials needed and used
- Information on the time and human resources required
- A list of equipment used
- The total cost of the work performed

A work order system should be designed to be functional rather than just additional paperwork. A sample work order is shown in Figure 10-4.

Technical Information

Water distribution systems are often swamped with technical information—information accompanying valves, hydrants, pumps, pipe, tapping machines, backhoes, and other distribution system components and equipment. Today's competitive sales market further adds to the volume of material by providing

WORK ORDER

Work Order Number: [WorkOrderID]

In Service: ________ **Out of Service:** ________

WO Initiated Date: [InitiateDate]	**Supervisor:**	[Supervisor]	**Team Members:**	[EstLaborName]	Shop: [Shop]

Activity Performed: [Description]	
Address: [Location]	**Key Map:** [MapPage]

Crew Type: WR SR SS FH MR VS VR EV PM **Location of Worksite:** Front Rear Side Arrival: ________ Departure: ________

Size Main: ¾" 1" 1¼" 1½" 2" 4" 6" 8" 10" 12" Other ______ **Size Service Connection:** ¾" 1" 1¼" 1½" 2" 4" 6" 8" Other ______

Pipe Type: A.C. C.I. Clay Concrete PVC Steel Other ______ **Material (quick codes) used:** ________________

Further Work Pending: Yes No Landscape Area: _____ Concrete Cut: _____ Asphalt Cut: _____ Fence: _____ wood chain link wrought iron brick other _____

Vehicle (quick code): ________ **Equipment (quick code):** ________ **Time Updated to C.O.S.:** ________ **Work Order Status:** [Status]

Update Map: [UpdateMap]

Instructions:	[Instructions]

Comments:	[Comments]

Associated Service Requests

RequestID	**Date/Time Initialized**	**Priority**	**Description**	**Problem Address**
[SRequestID]	[SRDateTimeInit]	[SRPriority]	[SRDescription]	[ProbAddress]
[associatedsr]				

Tasks

Order	**Name**	**Start**	**Status**	**Proceed**	**Assigned To**	**Request Code**	**Comments**
[TaskSeqID]	[TaskName]	[StartDate]	[TaskStatus]	[Proceed OK]	[AssignedTo Name]	[TaskCode]	[TaskComments]
[tasktable]							

Employee #s ______ ______ ______ ______ ______ **Planner/Scheduler Sign Out:** ________________

Figure 10-4 Sample work order form

Courtesy of Department of Public Works and Engineering, City of Houston.

technical information through the mail. Technical information should not be discarded simply because of its bulk. Many technical bulletins and pamphlets are extremely important because they provide specifications for installation, operating tips, or maintenance procedures.

Technical information should be evaluated and maintained on the computer for easy reference. Each major item within the file system should have its own section. For example, a section on hydrants could be established and subdivided by manufacturer. Files on each hydrant manufacturer would specify where important information pertaining to the hydrant is stored. They would also contain any documents, such as warranties, letters, or manufacturers' phone numbers. A filing system might also be established on valves, subdivided by valve type and manufacturer.

File systems can be arranged in any number of ways. The important consideration is that every water supplier should establish an efficient means of organizing important technical information for future reference.

The date received should be stamped or written on new catalogs or pieces of literature as it is received. Manufacturers often do not date their material, and when several different catalogs end up in a file, there is no way of telling which is most current unless they have been dated.

Computerized Maintenance Management

A computerized maintenance management system (MMS) is an organized way for a utility to keep track of its maintenance needs. An MMS has a wide range of features, including work order management, inventory management, purchase requisitioning, labor planning and time management, and analytic or predictive reporting. Other application systems that are not part of an MMS but can, with customization, be integrated include general ledger, accounts payable, accounts receivable, payroll, geographic information, automated mapping and facility management, and engineering sciences or modeling programs. Other, less typical applications are also possible.

Maintenance generates vast amounts of information and a great deal of data that must be managed in some way. All of these conditions make the maintenance department an ideal candidate for computerized information and work management systems. For example, a computerized tracking mechanism can estimate the time and materials needed for repairs based on such factors as reported symptoms, historic data on similar repair types, and previous work management information. Data listings show how much work is waiting to be done, how many jobs are currently in progress, how much work is completed and is waiting for final check or approval, and how many repairs are waiting for parts to arrive.

An effective maintenance system generates several kinds of work orders: preventive, corrective, emergency, and blanket. It allows for a routine scheduling of a known work backlog and the interrupt scheduling of nonroutine work that has a higher urgency.

Managers need an MMS that allows them to prioritize all work so that it can be completed in a timely and commonsense manner. When prioritizing work, managers should consider (1) whether the task is critical to system operation and (2) whether the task is critical to the organization. Many tasks, although not critical to continuous operation, may still be time-sensitive or highly valuable to the organization (e.g., housekeeping or safety tasks). The planning and scheduling procedures of a maintenance organization are largely dependent on the priorities of the tasks that need to be completed.

maintenance management system (MMS)
An organized, typically computerized, way for a utility to keep track of its maintenance needs.

Work can be scheduled in a number of ways. The system can monitor work requests with higher priorities and insert tasks into schedules automatically, or the maintenance manager can adjust schedules based on predetermined priority codes as he or she sees fit.

Formal requests for emergency work are usually entered into the system after the fact. Data regarding the nature of the repair, labor effort expended, and materials used are entered into the system. Including emergency repair information allows for a complete historical picture of equipment maintenance.

Other important elements of an MMS are described in the following paragraphs.

Reporting

The maintenance system must keep track of labor and material use. These data can help determine correct staffing levels or identify frequently used materials or parts. Reports should cover user-specified time periods and reflect data such as labor hours and cost (both regular and overtime), types of repairs, parts and materials used, and the uses and costs of outside contractors. These reports can be customized to provide executive reporting, which provides specific indicators tailored for overview by management.

Preventive Maintenance

Preventive maintenance (PM) is a significant function of automated maintenance systems. Effective PM means less time spent on corrective maintenance, resulting in better planning and increased productivity. The reliability of equipment can also be substantially improved.

The consistent, timely completion of regular PM tasks involves orderly procedures, schedules, and controls. The respective tasks to be performed on each piece of equipment should be identified and entered into the system. The usual sources of information are operational and maintenance manuals, equipment manufacturers' recommendations, and the experience of the plant personnel.

Tasks are categorized according to frequency of performance (e.g., daily, weekly, monthly, quarterly, per run-time hours, or by specific calendar date). A detailed procedure is defined for each task. This approach assists with new employee training. It also ensures uniformity (a standard operating procedure) in the performance of each task.

Equipment Histories

A maintenance history records all corrective and PM tasks performed on a particular piece of equipment for a specified period of time (1–5 years). It is vital for making repair-versus-replacement decisions.

Historical reports also include information vital to the analysis of equipment effectiveness. Data such as parts and materials used, personnel who performed the work, duration of tasks, reasons for equipment failure, outside contractors used, or environmental conditions can be important when trends or statistics need to be determined.

System Integration

An MMS should not be installed to run independently. System integration is important.

Connection points between maintenance systems can be divided into two categories: those that solely provide input to maintenance and those that also require output from maintenance. Repair notification can come from any interfacing departments. Any existing SCADA system can be interfaced to provide equipment run times for maintenance. This information, in turn, can automatically trigger appropriate PM tasks.

The operations department can provide information to the maintenance department regarding the intended operating plans. As a result, major repairs and PM tasks can be performed during scheduled shutdown periods.

The accounting department can provide budgetary information so that dollar amounts and limits can be established and monitored by account number.

The personnel department can establish codes for specialized skills and training of individual employees. It can update the system when new training and certification are acquired by each individual. This will allow for the scheduling of staff who possess the proper experience to do the job. The department can also provide labor rates and update accordingly so that the cost of each work order can be calculated.

The purchasing department can provide valuable vendor information regarding suppliers. It can monitor vendor performance, take advantage of cost-saving opportunities, and obtain material in a timely fashion. Purchase order status should be accessible to the maintenance department so that the expected delivery date of parts can be incorporated into the planning process. Maintenance also needs to be notified when parts are received.

Maintenance locators require information and maps from the geographic information system to mark infrastructure repair sites accurately. The engineering department can provide current, as-built drawings of equipment, piping, and wiring to assist with many repair tasks. Maintenance can monitor the progress of construction projects being conducted by the engineering and planning departments. This way, maintenance can be prepared to assume its activities when the projects are completed. New equipment and parts descriptions should be transferred to the maintenance database from project files before the equipment is put into use. This allows appropriate PM routines and procedures to be identified and incorporated.

Stores (warehouse) inventory provides the on-hand balance of parts for work order planning. This balance permits scheduling of only those work orders that have all of the necessary parts to complete the task. Inventory balances should automatically update the status of work orders. When all reserved parts are in stock, the work order can be released for scheduling and execution. A history of parts usage and an analysis of that usage can be determined from the information within the stores' inventory records.

Labor costs, material costs, and work order details are available to the accounting department. This department will incorporate the information into its budgeting process and possibly charge fees to user departments, other agencies, or customers. Labor hours charged to work orders can be input to payroll for time keeping and check processing. The purchase of special items required by the maintenance department can be controlled by the purchasing department and validated by the system via a purchase requisition system that is closely integrated with the maintenance department.

Work order status on infrastructure-mapping tasks is accessible through the GIS for use by any user department. Work order status on infrastructure and meter repairs is also available to the customer information system (CIS), so that customer service representatives can respond to inquiries from the public. In addition, the work order status, labor and material costs, and other data needed for managerial reports are available to the approved employees. Work order status can also be linked to equipment in the laboratory information management system (LIMS), so that the lab can plan when equipment will be used and when routine or corrective repairs will be completed.

All of these interface points provide the basis for a complete and comprehensive system integration plan. They allow for the accountability of each department's unique data while providing for a cooperative, shared information environment.

Other Major Information Systems

In addition to SCADA, AM/FM/GIS, and MMS, other major information systems are used by a water utility to manage the water treatment and distribution network. All of these systems can be connected together (i.e., interfaced) to share common data and applications. This section discusses the other major computer information system interfaces, along with the corresponding shared data and applications.

Source-of-Supply Systems and Treatment Plant Process Control Systems

The source-of-supply system is concerned with the quantity and quality of raw water. The treatment plant process control system is concerned with the quantity and quality of water being treated. Maintaining the quality and quantity of water is a function of the demand. Computer systems can handle the large amounts of data from historical records and from real-time, or current, network sensors.

How these two systems work with the other systems depends on how the system network as a whole is managed. If the network is automated but does not include demand forecasts, the impact of consumption on the network is passed on to the treatment facility, which in turn passes this impact on to the supply source. Network demand variations are partially smoothed out by the use of treated-water reservoirs, which reduce the need for flow variations at the treatment plant.

If the network is driven by an optimized system that takes into account demand forecasts, these forecasts are passed directly on to both treatment facilities and supply sources. These forecasts improve the utility's ability to keep a constant flow at the treatment facilities.

Laboratory Information Management Systems

A laboratory information management system collects, stores, and processes water quality information. It also gives feedback to enhance water treatment and generates reports.

Information exchanges with an LIMS are limited to network quality management data. The control center receives real-time quality data from the LIMS. These data are used as they are received, especially in the form of alarms, but they are also stored in a water quality database, as are other analyses coming from the laboratory. These data can then be used to simulate water quality changes in the network, in connection with data on pump operation and valve status. Conversely, when the laboratory detects an abnormal analysis, the data are transmitted to the network manager so that corrective measures can be taken immediately.

Leakage Control and Emergency Response

Leak detection software can be used to augment traditional approaches to leak detection. For instance, the software can receive readings of pressures, reservoir levels, and flow measurements in order to send input to a hydraulic model.

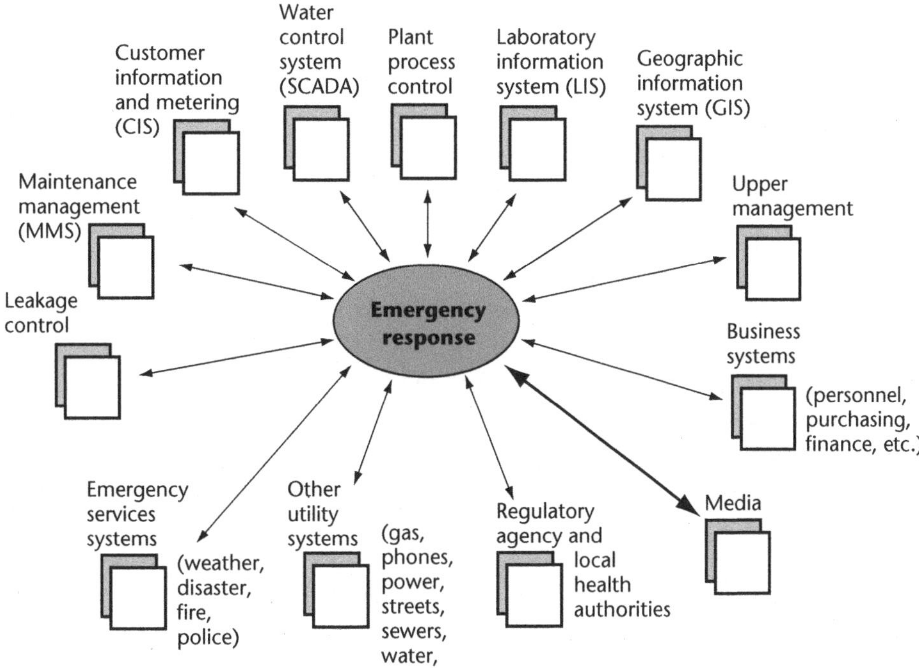

Figure 10-5 Emergency response communications

Alternatively, the software can receive meter readings from the distribution network to calculate consumption per district. Unusual readings can indicate a leak.

Emergency response information systems can also benefit from the integration of computerized systems. Figure 10-5 illustrates how shared data from inside the water utility, as well as from outside sources including emergency services and other utilities, are available in an emergency situation. Having all of the required information readily available allows a water utility to react much more quickly in an emergency.

Customer Information Systems

The CIS contains data such as consumption and the customer's category and address. This information is used to calculate the meshed network for demand forecasts and to calculate network yields. Remote or automatic meter-reading methods provide data that can be used for demand forecasting and for leak detection.

Study Questions

1. Data manipulations using the basic elements of an automatic mapping/facility mapping/geographic information system (AM/FM/GIS) would include
   a. collection and storage.
   b. collection and management.
   c. modeling and analysis.
   d. retrieval, display, and storage.

2. Which data category for water utility data sets includes the following map layers: control information, planimetric features, and hydrology features?
   a. Land records data
   b. Natural resources data
   c. Watershed resources data
   d. Base data

3. A _____ is used to communicate field information back to the main office and to order materials, organize field crews, and obtain easements?
   a. material safety data sheet
   b. facility memorandum
   c. work order
   d. record of activity

4. Which data management system is best suited to activities that are not time sensitive or mission critical?
   a. Geographic information system
   b. Automated mapping/facility management system
   c. Decision support system
   d. Operational support system

5. In a typical water utility data set, property boundaries, easements, and right-of-ways are examples of
   a. base data.
   b. land records data.
   c. facilities and distribution systems data.
   d. transportation network data.

6. What is the general term for an organized, typically computerized, way for a utility to keep track of its maintenance needs?
   a. Customer information system
   b. Laboratory information management system
   c. Maintenance management system
   d. Geographic information system

7. What are the main elements of most AM/FM/GIS databases?

8. What is the common element of AM/FM/GIS databases that integrate all of the information?

9. What are the two types of maintenance activities common to computerized maintenance management systems?

10. What type of data management system collects, stores, and processes water quality information?

Chapter 11
Safety, Security, and Emergency Response

Water Supply System Threats

The types of threats that would affect water utilities include both natural and accidental disasters and intentional disasters. Natural disasters include floods, windstorms, ice storms, snowstorms, fires, droughts, and earthquakes. Accidental disasters include chemical spills, fires, transportation accidents, and explosions. Intentional disasters include the terrorist activities that are covered in this chapter. Three main types of consequences may occur: (1) complete interruption of supply, (2) supply of sufficient quantity, but compromised quality, and (3) supply of sufficient quality, but insufficient quantity.

Water distribution systems are extensive, with many components and subcomponents. The components and subcomponents are relatively unprotected and accessible and are often isolated. The physical destruction of a water distribution system's components and subcomponents or the disruption of water supply could be more likely than the introduction of contaminants to a system. The actual probability of a terrorist threat to drinking water is probably very low; however, the consequences could be extremely severe for exposed populations.

Four major types of intentional threats to drinking water systems (Mays, 2004) are discussed here.

Cyber Threats

The electronic control functions intended to make water treatment more efficient and effective can be turned against the treatment system when hacked or otherwise violated. Examples of cyber threats include the following:

- Physical disruption of a supervisory control and data acquisition (SCADA) network
- Attacks on the central control system to create simultaneous failures
- Software attacks using worms/viruses
- Network flooding
- Jamming of control system functions
- Data corruption to give the appearance of appropriate chlorination when in fact no disinfectant has been added, allowing the proliferation of microbes

Physical Threats

Physical destruction of a system's assets or disruption of water supply is perhaps more likely than is contamination of the water supply. A single terrorist or a small group of terrorists could easily cripple an entire city by destroying a critical component of the water system. For example, opening and closing major control valves and turning pumps off and on too quickly could create a water hammer effect and, consequently, simultaneous main breaks. The resulting loss of water pressure would compromise firefighting capabilities and possibly lead bacterial buildup in the system.

Chemical Threats

There are numerous chemical warfare agents and industrial chemical poisons. Some of the chemical warfare agents include hydrogen cyanide, tabun, sarin, VX, lewisite (arsenic fraction), sulfur mustard, 3-quinuclidinyl benzilate, and lysergic acid diethylamide. Some of the industrial chemical poisons include cyanides, arsenic fluoride, cadmium, mercury, dieldrin, sodium fluoroacetate, and parathion.

Biological Threats

Several pathogens and biotoxins exist that have been weaponized, are potentially resistant to disinfection by chlorination, and are stable for relatively long periods in water. The pathogens include *Clostridium perfringens*, plague, and others. Biotoxins include botulinum, aflatoxin, ricin, and others. Water does provide dilution potential; however a neutrally buoyant particle of any size could be used to disperse pathogens into drinking water systems. Water storage and distribution systems can facilitate the delivery of an effective dose of toxicant to a potentially large population. A more extensive discussion of microbiological contaminants and threats of concern is presented by Abbaszadegan and Alum (2004).

Water Supply System Vulnerabilities

Vulnerabilities of Water Supply Systems

Water supply systems are designed to operate under pressure and supply most of the water for firefighting purposes. Either a loss of water or a loss of pressure could interrupt service, disable firefighting ability, and disrupt public confidence. Such loss of pressure and/or water could result from sabotaging pumps that maintain flow and pressure or from disabling electric power sources, causing long-term disruption and taking months to replace custom-designed equipment.

Locations of vulnerability in these systems include the following (Clark and Deininger, 2000):

- Raw-water source (surface or groundwater)
- Raw-water channels and pipelines
- Raw-water reservoirs
- Treatment facilities
- Connections to the distribution system
- Pump stations and valves
- Finished water tanks and reservoirs

Figure 11-1 shows a range of potential contamination scenarios.

Vulnerabilities of Computer System Infrastructure

The computer infrastructure of a medium- to large-sized water utility has numerous systems, including the financial, human resources, laboratory information management, maintenance management, supervisory control and data acquisition (SCADA), and others. The term SCADA is often used to include both in-plant computer-based process control systems and computer-based systems providing monitoring and control of geographically distributed raw-water production and

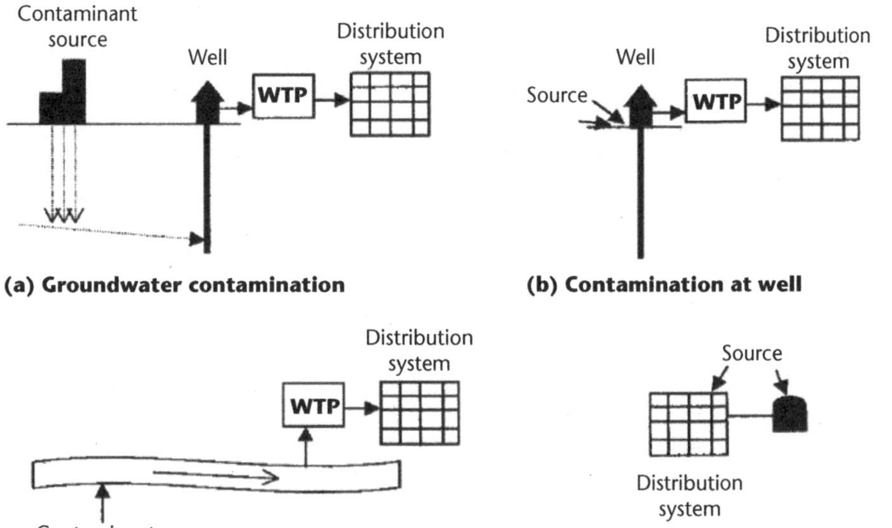

Figure 11-1 Contamination scenarios
Source: Grayman et al., 2004a.

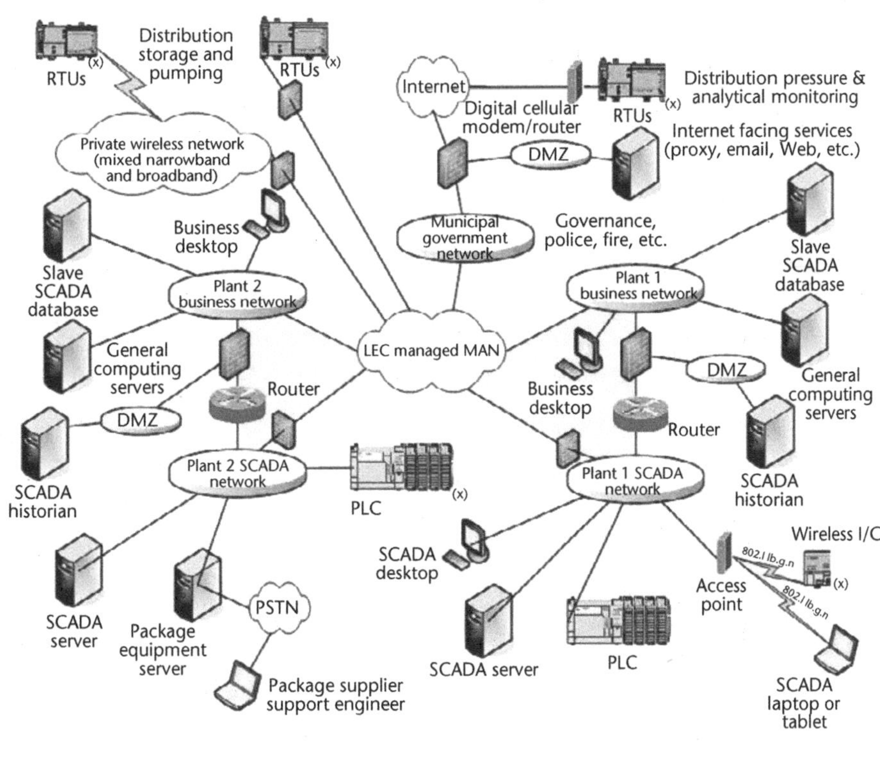

Figure 11-2 Typical information flow requirements for medium- to large-sized water utilities
Source: Phillips, 2009.

treated water distribution systems. Figure 11-2 simplifies a utility's network architecture into blocks showing the composite of a network's vulnerabilities.

The list of vulnerabilities determined from a USEPA-mandated assessment of SCADA systems of a number of large utilities in the United States are as follows (Panguluri et al. 2004):

- Operator station logged on all the time even when the operator is not present at the workstation, thereby rendering the authentication process useless
- Relatively easy physical access to the SCADA equipment

- Unprotected SCADA network access from remote locations via digital subscriber lines (DSLs) and/or dial-up modem lines
- Insecure wireless access points on the network
- Most of SCADA networks directly or indirectly connected to the Internet
- No firewall installed or the firewall configuration is weak or unverified
- System event logs not monitored
- Intrusion detection systems not used
- Operating and SCADA system software patches not routinely applied
- Network and/or router configuration insecure; passwords not changed from manufacturer's default

Vulnerability Assessments

A vulnerability assessment is the process of identifying, quantifying, and prioritizing (or ranking) the vulnerabilities in a system. Vulnerability assessments help water utilities to evaluate their susceptibility to potential threats and identify corrective actions to reduce or mitigate the risk of serious consequences from vandalism, insider sabotage, or terrorist attack. As required under the Bioterrorism Act (Public Health, Security, and Bioterrorism Preparedness and Response Act, PL 107-188, June 2002), a drinking water utility serving more than 3,300 persons must do the following:

- Conduct a vulnerability assessment, certify to USEPA that the assessment has been completed, and submit a copy of the assessment to USEPA.
- Show that the system has updated or completed an emergency response plan outlining response measures if an incident occurs.
- Certify to the USEPA administrator, upon completing the vulnerability assessment, that the system has completed or updated its emergency response plan.

Vulnerability assessments for a water system share the following common elements (USEPA, 2002):

- Characterization of the water system including its mission and objectives
- Identification and prioritization of adverse consequences to avoid
- Determination of critical assets that might be subject to malevolent acts that could result in undesired consequences
- Assessment of the likelihood (qualitative probability) of such malevolent acts from adversaries
- Evaluation of existing countermeasures
- Analysis of current risk and development of a prioritized plan for risk analysis

The complexity of vulnerability assessments ranges based on the design and operation of the system.

SCADA Systems

A 21-step guide to improve cyber security of SCADA networks was developed by the US Department of Energy (2002) listed below:

1. Identify all connections to SCADA networks.
2. Remove unnecessary connections to the SCADA network.

3. Evaluate and strengthen the security of any remaining connections to the SCADA network.
4. Harden SCADA networks by removing or disabling unnecessary services.
5. Do not rely on proprietary protocols to protect the system.
6. Implement security features provided by device and system vendors.
7. Establish strong controls over any medium that is used as a backdoor into the SCADA network.
8. Implement internal and external intrusion detection systems and establish 24-hour-a-day incident monitoring.
9. Perform technical audits of SCADA devices and networks, and any other connected networks, to identify security concerns.
10. Conduct physical security surveys and assess all remote sites connected to the SCADA network to evaluate their security.
11. Establish SCADA "Red Teams" to identify and evaluate possible attack scenarios.
12. Clearly define cyber security roles, responsibilities, and authorities for managers, system administrators, and users.
13. Document network architecture and identify systems that serve critical functions or contain sensitive information that require additional levels of protection.
14. Establish a rigorous, ongoing risk management process.
15. Establish a network protection strategy based on the principle of defense in depth.
16. Clearly identify cyber security requirements.
17. Establish effective configuration management processes.
18. Conduct routine self-assessments.
19. Establish system backups and disaster recovery plans.
20. Senior organizational leadership must establish expectations for cyber security performance and hold individuals accountable for their performance.
21. Establish policies and conduct training to minimize the likelihood that organizational personnel will inadvertently disclose sensitive information regarding the SCADA system design, operations, or security controls.

Early Warning Systems

Early warning systems (EWSs) for both source and finish (distributed) water are intended to reliably identify low-probability, high-impact contamination events. Figure 11-3 illustrates the elements of a source early warning system. The International Life Sciences Institute (ILSI, 1999) identified the following objectives for hazardous events in water:

- Provide warning in sufficient time to respond to a contamination event and prevent exposure of the public to the contaminant
- Have the capability to detect all potential contamination threats
- Can be operated remotely
- Can identify the point at which the contaminant was introduced
- Have a low rate of false-positive and false-negative results
- Provide continuous, year-round surveillance

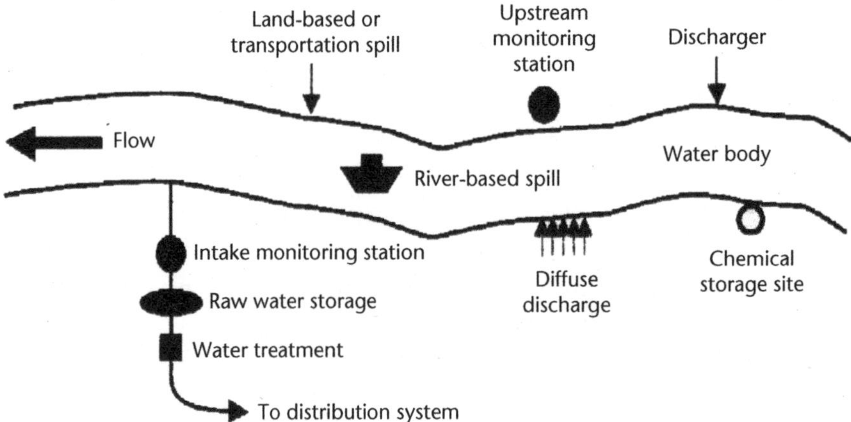

Figure 11-3 Elements of an early warning system
Source: Grayman et al., 2004b.

- Produce results with acceptable accuracy and precision
- Require low skill and training
- Is affordable to the majority of public water systems

Keeping in mind that a contamination event in a source or in a distributed water system must be identified in time to allow for an appropriate response that mitigates or eliminates the adverse impact, the following are features of an ideal EWS (Clark et al., 2004):

- Has a rapid response time
- Is fully automated
- Screens for a range of contaminants
- Is specific for the contaminants of concern
- Offers sufficient sensitivity
- Has a low occurrence of false positives and false negatives
- Has a high rate of sampling
- Is reliable and rugged
- Requires minimal skill and training
- Is affordable
- Maintains an online model that is ready to use for source tracing after a contamination event

Security Hardware and Surveillance Systems

Intrusion Detection Systems

An intrusion detection system (IDS) has the purpose of detecting an intruder approaching a site, facility, or area as early as possible. These systems may provide the first and only indication that someone or something is trying to enter a facility. Key issues to consider in designing an IDS are the following (Booth et al., 2004):

- What is the facility, space, or area to be protected?
- Who is the perceived threat?
- What are known vulnerabilities of the area or space?
- What are the key assets or targets at the space?

- How will system monitoring take place?
- What power and communications methods exist?

The design of a perimeter IDS must satisfy the following concepts, with the goal of achieving the best possible performance (Booth et al. 2004):

- *No gaps in coverage.* The IDS must provide a continuous line of detection around the perimeter area or interior space.
- *Suitability for physical and environmental conditions.* The sensor must be appropriate for the area being monitored (temperature, humidity, rain, fog, wind, pollution).
- *Layers of protection.* A fundamental security concept is that multiple, layered detection systems are much more effective than single systems. If one system is bypassed or defeated, the remaining systems are still in place to detect the intruder.

There are several intrusion detection sensor categories, including the following:

- *Exterior intrusion sensors* are used to sense an intrusion crossing an outdoor perimeter boundary. There are several types of exterior intrusion detection sensors, including buried line sensors, pressure/seismic sensors, magnetic field sensors, ported-coaxial buried cable systems, and fiber-optic buried cable systems.
- *Fence-mounted cabling sensors* include electromechanical vibrating sensing, coaxial strain sensitive cable, fiber-optic strain sensitive cable, taut-wire systems, and fence-mounted electric field sensors.
- *Free-standing exterior sensors* include active infrared sensors, passive infrared sensors, microwave, and dual technology (passive infrared [PIR] and microwave).
- *Interior sensors* include interior volumetric sensors, ultrasonic and microwave sensors, passive infrared, and dual-technology sensors.
- *Interior boundary penetration sensors* include glass break sensor, door switch, and linear beam (photoelectric beam).

Digital motion detection analyzes video streams of closed-circuit television cameras and compares those video streams to a still image in the unit's memory. The cameras must be in a fixed position and not pan- or tilt-type units. Wireless intrusion detection sensors include door switches, PIR, and dual technology (microwave plus PIR) sensors. Wireless sensors can provide cost savings over hardwired sensors.

General intrusion detection recommendations for water supply systems are (Booth et al. 2004):

- Reservoirs and elevated tanks
  - Perimeter detection
  - Monitoring ladder
  - Monitoring vaults
  - Monitoring hatches
- Pump stations
  - Perimeter detection
  - Entrance detection
  - Volumetric detection

- Water treatment stations
  - Perimeter detection
  - Entrance detection
  - Volumetric detection
  - Monitoring clearwells
  - Monitoring SCADA control rooms
  - Monitoring chemical storage and dispensing areas
- Raw-water intake stations
  - Perimeter detection
  - Entrance detection
  - Volumetric detection

Closed-Circuit Television

Closed-circuit television (CCTV) has been used for decades as an integral part of comprehensive security systems. The evolution in digital hardware has made CCTV smarter, more reliable, more efficient, and more effective for premise security in all types of applications (Booth et al., 2004). CCTV system components include cameras, the switcher or multiplexer, the transport media, and the wireless transmission.

Access Control Systems

Access control systems permit only authorized personnel to enter and exit a restricted area. Electronic access control systems are used to control entry into a perimeter, area, or interior space. Layered security systems for a water supply system may include four or five security access control levels (Booth et al., 2004):

- Public zone (level 1)
- Clear zone (level 2)
- Building lobby area (level 3)
- Internal circulation area (level 4)
- High-value areas, if needed (level 5): for example, SCADA rooms, security equipment rooms, laboratory areas, chemical storage, etc.

Locking systems include key locks, mechanical or electrical keypads, electrified locking systems, fail-safe locking systems, and fail-secure locking systems. Card reader systems provide the most reliable, flexible method of controlling access to a facility. The access control for the interior circulation area could be a card access, and the access control to the high-value areas could be a card reader plus a personal identification number. Figure 11-4 illustrates a typical single-door card reader installation, and Figure 11-5 illustrates a typical access card system block diagram (with video surveillance). A guard tour system requires that designated security staff conduct a security tour of a facility at specified frequencies and durations.

Role of SCADA Systems for Security

SCADA systems can be made a central part of security efforts so that security measures are coordinated with operations. Linking SCADA systems to perimeter monitoring devices provides constant monitoring and reduces the need for manned patrols. Security systems and equipment can be interfaced directly to the SCADA system or through a remote terminal unit (RTU). The SCADA

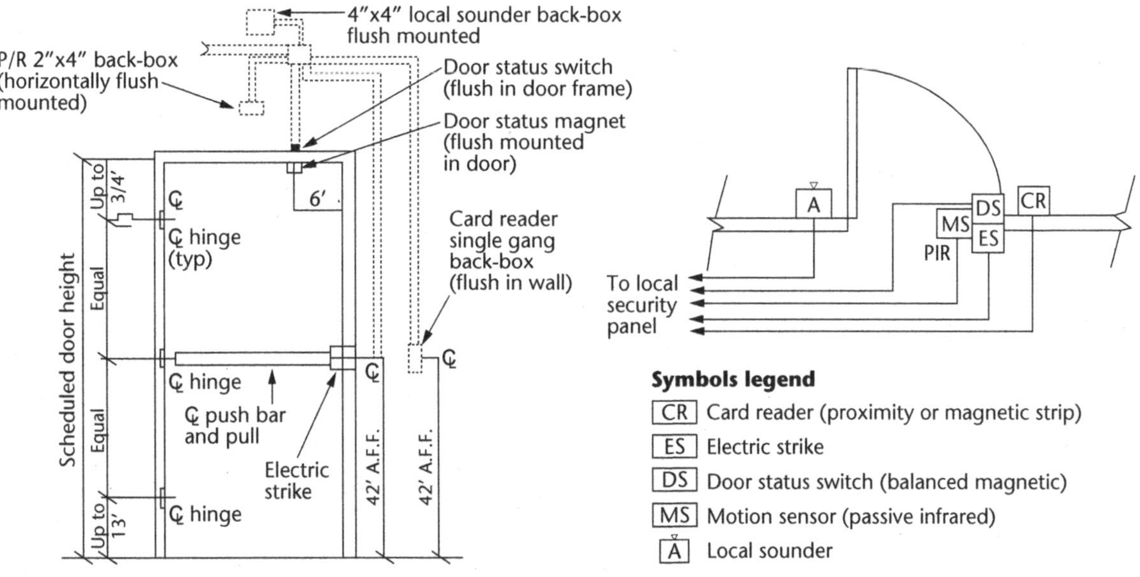

Functionality:
In the normal state the door is closed, locked on the outside and unlocked on the inside. Normal operation is by card reader on the outside and by motion detector on the inside. The motion detector will shunt an alarm and enable authorized exit. If the door is opened without the operation of the card reader or motion detector, the local sounder will sound. The alarm condition is monitored at the security terminal.

Door detail (single door with card reader, local sounder, and electric strike)

Figure 11-4 Typical single-door card reader installation

Courtesy of Booth et al., 2004.

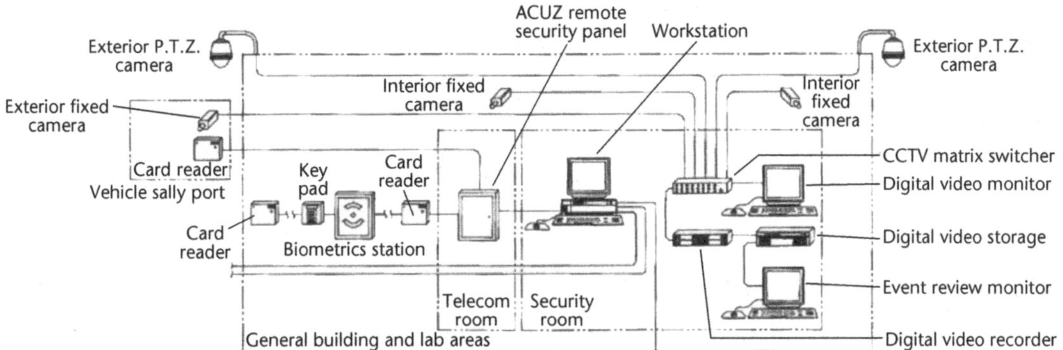

Figure 11-5 Typical access card system block diagram (with video surveillance)

Courtesy of Booth et al., 2004.

system can react to conditions and perform control actions automatically. These actions could include the starting and stopping of pumps, the opening and closing of valves, emergency shutdowns, and others. Portions of a water distribution system can be isolated by stopping pumps and closing valves. The SCADA system can also include alarm management. SCADA systems can coordinate security measures with process operations; reduce or eliminate manned patrols; provide constant monitoring, system-wide; and record alarms and events. SCADA systems can be expanded by using additional input/output points, RTU devices, and network links. SCADA systems can be enhanced even further if they incorporate advanced capabilities such as hydraulic modeling and simulation (network analysis).

Emergency Response Plans

An emergency response plan (ERP) provides a step-by-step response to, and recovery from, incidents related to emergencies. The ability of water utility staff to respond rapidly in an emergency will help prevent unnecessary complications and protect consumers' health and safety. Proper preparedness is the key to achieving emergency response success.

System components could become less susceptible to harm by taking mitigation measures, which are actions that are taken to eliminate or reduce the harmful effects of water systems emergencies. The following lists provide a guide to mitigation measures for water distribution, which should be described in an ERP (Manitoba Office of Drinking Water, 2009):

Mitigation at the raw-water source includes the following measures:

- Having access to an alternate raw-water source, if situation allows
- Restricting access of unauthorized persons by fence and gate
- Facilitating access to the water source by utility staff (access by boat, road)
- Maintaining wells and surface water intakes; applying setback distance to wellhead
- Having a source water protection plan and wellhead protection plan

Mitigation of water distribution system failures consists of the following measures:

- Having spare parts available (valves, pipes, repair kits)
- Maintaining networks by replacing old, damaged, and poorly built distribution system components; regular flushing, valve and hydrant exercising
- Having redundancy by close-looping of networks and installing sufficient check valves, other control valves, etc.
- Preparing/updating distribution network mapping

Components of an ERP

The components of an ERP could include the following (Manitoba Office of Drinking Water, 2009):

- A detailed map of the distribution system; detailed locations of each valve in the system, including references that will aid in locating these valves; and a map of well locations and surface water intakes, as applicable
- A detailed map of electrical diagrams clearly showing generator and power source change-over
- A contact list of emergency services, regulators, suppliers, contractors, water users with critical needs, media, phone companies, and water utilities
- A statement of amounts budgeted for emergency use, along with a statement showing who may authorize expenditures for such purpose and under what conditions
- A determination of not less than nine most likely emergencies that may affect the water system and procedures to be followed and actions necessary to provide service during emergencies
- A determination of who would operate the system if all operators are off (e.g., pandemic flu)

- A description of ways to obtain and transport water from an alternate source, should it become necessary (It is advisable to have arrangements for obtaining water from at least two alternative sources that are not likely to be affected by the same hazards at the same time.)
- A description of how often the plan should be revised (e.g., at least every 2 years), who has copies of the plan, and other logistical information relating to the plan
- A description of methods for notifying water users that an emergency is under way

An example ERP and an example emergency action chart are shown in Figures 11-6 and 11-7, respectively.

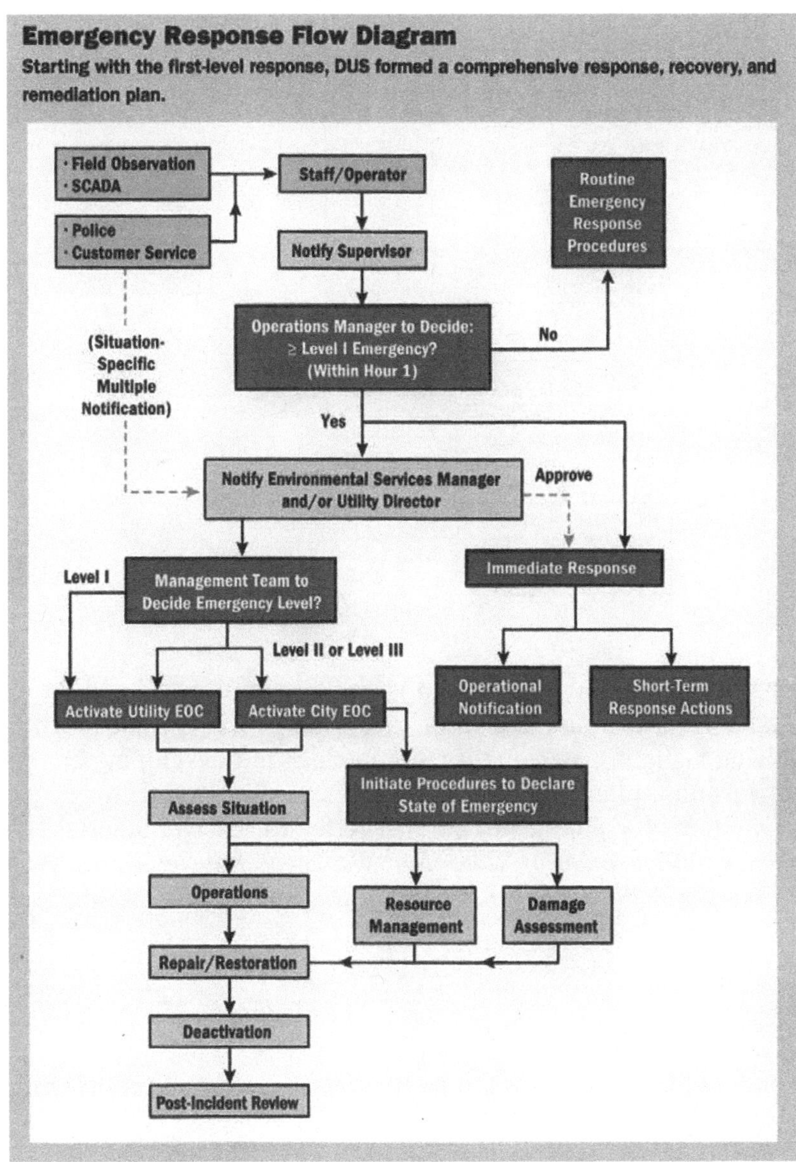

Figure 11-6 Emergency response flow diagram for Henderson, Nevada
Courtesy of Johnson and Gabriel, 2009.

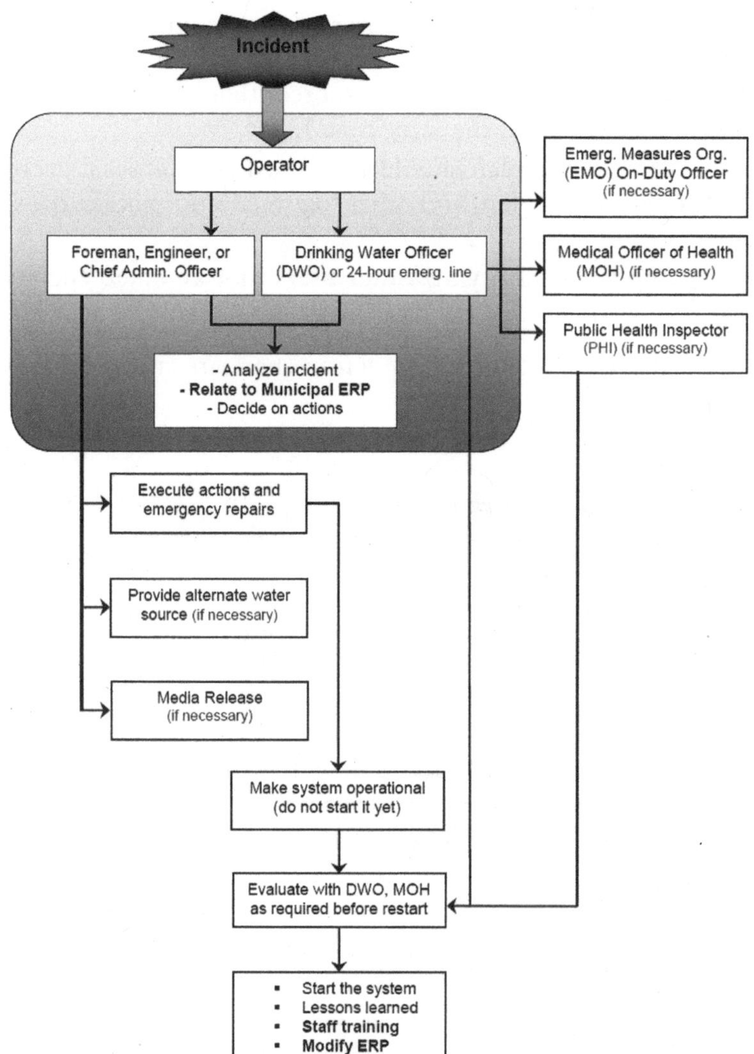

Figure 11-7 Emergency action chart

Courtesy of Manitoba Office of Drinking Water, 2009.

Although the potential threats to a water distribution system are many, utilities can take steps to minimize risks by taking proper physical/cyber security steps, applying vigilant water quality monitoring, and developing an appropriate emergency response plan. Training and practice are essential to maintain workforce preparedness for emergency events. Refer to AWWA Standard J100 *Risk and Resilience Management of Water and Wastewater Systems* and AWWA Manual M19 *Emergency Planning for Water Utilities* for additional information.

Study Questions

1. Which of the following is a potential biological biotoxin threat?
   a. Sarin
   b. Ricin
   c. Taban
   d. 3-quinuclidinyl benzilate

2. The IDLH (immediately dangerous to life and health) value for chlorine, set by the National Institution of Occupational Safety and Health, is
   a. 10 ppm.
   b. 15 ppm.
   c. 25 ppm.
   d. 30 ppm.

3. How does hydrogen sulfide gas kill a person?
   a. Causes asphyxiation
   b. Paralyzes the respiratory system
   c. Paralyzes the nervous system
   d. Stops the heart

4. If sodium hypochlorite comes into contact with the skin, it should be immediately flushed with water for at least
   a. 10 minutes.
   b. 15 minutes.
   c. 20 minutes.
   d. 30 minutes.

5. A(n) _____ provides a step-by-step response to, and recovery from, incidents related to emergencies.
   a. emergency response plan
   b. vulnerability assessment
   c. intrusion detection form
   d. all-hazards plan

6. What are the four major types of intentional threats to drinking water systems?

7. What is the term for the process of identifying, quantifying, and prioritizing (or ranking) the vulnerabilities in a system?

8. Which agency developed a 21-step guide to improve cyber security of SCADA networks?

9. What are the three chief requirements of an intrusion detection system?

Chapter 12
Administration and Public Relations

Managing System Operations

The objective of distribution system operations is to reliably and efficiently deliver high-quality drinking water to meet customers' needs and satisfy fire flow requirements. Achieving this objective can be difficult. To be successful, system operators must understand the purpose and consequence of the many operational options they can employ. Then they must develop an integrated operations management plan that can be implemented and the results tracked.

Three main distribution system performance categories are included in the distribution system operations objective: water quality, reliability, and efficiency. Achieving excellence in one category may adversely affect the performance in the other two. System operators must balance these sometimes-competing objectives to satisfy the priorities established by their customers, regulatory agencies, and system managers. These categories are further defined as follows:

- *Water quality.* Ideally, water delivered to customers would be the same quality as the water that entered the distribution system (from the treatment plant or other source). However, water quality is usually changed as it travels through miles of distribution piping and resides in storage facilities. These changes mostly do not improve the water quality. System operators can use many practices to mitigate the changes and deliver water of high quality to users.
- *Reliability.* The goal of system operators is to deliver an uninterrupted supply of water to customers and to meet firefighting requirements. Main breaks, power supply problems, and emergency failures (from natural disasters or intentional acts) can cause temporary interruption to water service. System operators can adopt practices and plans to reduce the frequency and extent of these service stoppages.
- *Efficiency.* Energy usage and water losses are the two main elements affecting efficient delivery of water. Reducing these factors can save water and reduce distribution system operating costs. System operators can participate in improving efficiency by employing practices and procedures as part of an overall plan.

Many operational practices can affect the performance of the distribution system. System operators are often overwhelmed with the choices and resort to a reactive approach by using practices only when a problem is apparent. A reactive approach can be effective, but most often the effect is only temporary and is not documented, so when the problem reoccurs, the operators are again searching

operations management plan
A plan detailing the distribution system's performance objectives, operational practices to achieve the objectives, and performance measures and numeric goals of the plan.

for a solution. To ensure lasting improvement in system performance, operators should use a management plan that is based on a thorough assessment of the system and selection of the operational practices that can make a difference.

The management plan should include performance measures and goals for individual practices and the three main performance categories (water quality, reliability, and efficiency). System operators should continuously review the performance results so the plan can be modified if needed.

Some of the performance improvement practices are necessarily only initiated by executive management because they may involve considerable capital expense. However, operators must be knowledgeable of these initiatives so they can operate the system effectively if they are adopted. Operators should also be included for input by management when considering system performance-enhancing options. Thus, operators must be knowledgeable of the merits of these options so they can provide educated advice to management.

Developing a System Operations Management Plan

A distribution system operations management plan includes performance objectives, operational practices to achieve the objectives, and performance measures and numeric goals of the plan. Also, persons or departments responsible for leading implementation of certain operational practices are identified in the plan. Periodic plan reviews are scheduled to review the results and make modifications if needed.

System Performance Assessment

An initial step in developing an operations management plan is to assess the current status of the system performance. Evaluating performance for water quality, reliability, and efficiency can be an involved process. However, there are several good resources that can simplify this task.

The Partnership for Safe Water is supported by the six largest drinking water organizations, including the American Water Works Association (AWWA) and US Environmental Protection Agency (USEPA). The Partnership provides a comprehensive program to optimize distribution system performance. A key element of this program is a self-assessment of the current system status. The results of the self-assessment are then used to create an improvement plan that leads toward improved performance. The program provides extensive guidance on how to conduct the self-assessment. This includes software to calculate performance measures that are used to assess the degree of optimization. Performance goals (that define excellence) are provided so that systems can compare and track the results of their improvement action plan. AWWA offers further information about the program on its website (www.awwa.org).

AWWA has developed a standard for distribution system management, ANSI/AWWA G200. This standard includes the consensus management practices of the drinking water industry. Conformance with the standard is voluntary. A companion document for the standard is the *Operational Guide to AWWA Standard G200: Distribution Systems Operation and Management* (Oberoi, 2009). This guide provides useful information on how to use the standard, including conducting a system self-assessment.

Another reference that may be helpful is *Water Distribution System Assessment Workbook* (Smith, 2005). The workbook contains worksheets for conducting a self-assessment, and references are provided for additional information.

Performance Measures and Goals

An important element of an operations management plan is to use performance measures and track system results against established goals. The most appropriate measures are identified in the system performance assessment.

Several approaches are used when establishing goals. Goals can represent the ultimate performance level, can reflect current operation, can be short term or long term, can be those recommended by others or internally developed, and may include financial limitations. No matter which approach is used, the goals must be recognized by all operations staff.

Goals should be achievable with effort. One approach is to have an interim or short-term goal that is slightly better than the current status. This provides for interim success while allowing time to reach a higher level. Good goals are those that are within the control of the person or group that is responsible. There does not need to be a goal for everything being measured. Some performance measures are just for tracking and do not need a goal attached.

Operational Practices

The distribution system operations management plan lists the practices used to meet the system performance objectives. The main operational practices affecting performance are listed in Table 12-1. Systems may emphasize selected operational

Table 12-1 Operational practices affecting distribution system performance

	Performance		
Operational Practices	**Water Quality**	**Reliability**	**Efficiency**
Corrosion control—external		•	
Corrosion control—internal	••		•
Cross-connection control	••	•	
Customer complaint response	•	•	•
Energy management		•	••
Flushing program	••	•	•
Hydrant and valve maintenance	•	••	•
Main breaks management	•	••	•
Maintaining disinfectant residual	••	•	•
Nitrification control (chloramines)	•		
Disinfection by-product control	•	•	
Pipe rehabilitation and replacement		••	•
Pressure management	•	•	••
Sediment control	•		•
Security and surveillance	•	•	•
Storage tank management	••	•	•
Water age management	••	•	•
Water loss control	•	•	••
Water sampling and response	•	•	•

• Performance-affecting practice.
•• Major performance-affecting practice.

practices depending of the need and the performance objectives. Many systems may list *all* of the listed operational practices in their plan.

The operational practices listed in Table 12-1 are the most common. Site-specific conditions may require the use of additional practices to meet system performance objectives. Examples of performance measures commonly used by utility system operators are given in this chapter. The table indicates the performance criteria mainly affected by each operational practice. Exceptions to these relationships may occur because of unusual local conditions, but these practices should be included in the management plan.

Discussion regarding development of a management plan for pipeline rehabilitation and replacement is included in this chapter because these practices are often adopted by executive management and included in a capital improvement plan.

Management Plan Review and Revision

After the distribution system operations management plan has been developed and is in use, it should be periodically reviewed to evaluate the results. Most systems review the plan annually. This review occurs most often after the data from a calendar year are available and before budgets for the next year are due. For many systems, the best time is in the spring.

The review includes an assessment of the performance measures as compared to the established performance goals. Any major operational challenges are noted along with suggestions how these could be approached if they reoccur. Projections for operations during the upcoming year should identify any unusual situations that may affect system operating practices. Adjustments to the operations management plan are made when performance goals are not being met or when changing circumstances require a modified approach.

System Operations Training

Managing the operation of a distribution system requires a trained staff. Distribution system operators require certification to ensure a level of understanding needed to provide potable water service. System operators need to meet regulatory requirements and provide emergency service for firefighting and in response to natural disasters.

Training can be provided through several sources: experienced operators, reference materials, training courses, and system operations simulations. Experienced operators provide invaluable information to individuals new to the system. Reference materials (like this book) provide standardized information gained from the operating experience of industry leaders and researchers. Also, these materials are usually the basis for certification exam questions. AWWA and other providers offer periodic training courses aimed at system operators (contact your local AWWA for availability).

Technology is providing another valuable training tool: hydraulic model operations simulations. Network computer models can be used to experience operational situations and practice responses. The simulations can be based on fixed information or real-time operating data. Extreme events can be simulated so that operators are able to develop confidence when faced with catastrophic situations. If this technology is not available, as a substitute, operators can talk through various operational scenarios so that they can develop confident responses.

Example Performance Measures

Performance measures are given in the following list for the operational practices listed in Table 12-1. These measures are commonly used by utility system operators to quantify performance and to compare with established goals. Measures are not listed for all of the operational practices. In some cases, this is because no common measures could be found. Performance goals are also not listed because each system must establish its own according to its needs. Some common goals are available in particular where regulatory requirements exist. Many systems have adopted the regulatory limit as their goal. Other systems have adopted goals that are more stringent than regulations require. The goal-setting approach is specific to each system and its operational objectives.

Customer Complaint Response

- Technical complaint rate = number of technical complaints ÷ number of customer accounts.
- Technical complaint response time = minutes to respond after receipt.

Technical complaints are directly related to the core services of the utility. They include complaints associated with water quality, taste, odor, appearance, pressure, main breaks, and disruptions of water service.

Energy Management

- Electrical usage rate = kW·h ÷ number of customer accounts.
- Unit cost of water delivery = $ cost ÷ mgd, or $ cost ÷ number of customer accounts.

Hydrant and Valve Maintenance

- Percentage of valves inspected and exercised annually = number of inspected ÷ total number in system × 100.
- Percentage of hydrants inspected and exercised annually = number of inspected ÷ total number in system × 100.
- Percentage of Large valves inspected and exercised annually = number of >10-in. valves inspected ÷ total large valves × 100.
- Hydrant repair time = time (hours) returned to service upon receipt of repair order.

Main Breaks Management

- Main break frequency rate = number of reported breaks per 100 miles of utility-controlled mains.
- Main break frequency trend = a 5- or 10-year main break frequency rate.

Reported breaks are those leaks or breaks that come to the attention of the water utility as reported by customers, traffic authorities, or any outside party due to their visible and/or disruptive nature. A nonsurfacing break initially reported as loss of pressure by a customer is an example of a nonsurfacing reported leak. Events that can be inferred from alerts by SCADA systems can be labeled as reported breaks. Water utilities respond to reported breaks in a *reactive* mode, often under emergency conditions (AWWA Manual M36, *Water Audits and Loss Control Programs*).

Maintaining Disinfectant Residual

- System disinfectant residual = monthly 95th percentile of all routine measurements.
- Goal is no consecutive residual measurements below minimum at individual routine sample locations.

Routine measurements are measurements taken from samples that are collected on a regular schedule.

Pipeline Rehabilitation and Replacement

- Pipeline annual renewal rate = miles of pipeline renewed ÷ total miles of distribution system pipeline mains.
- Pipeline annual breaks rate = number of breaks ÷ 100 miles of pipeline mains.
- Service interruption rate = number of 4-hour service interruptions ÷ number of customer accounts.

Pressure Management

- Minimum pressure (in each zone)
  - Normal water demand period, monthly average of daily minimums
  - Peak water demand and fire flow periods, hourly minimum during simultaneous peak water demand and fire flow conditions
  - Emergency conditions, during emergencies such as main breaks and power outages

Storage Tank Management

- Storage facility sampling frequency = number of water quality samples taken monthly.
- Storage facility inspection frequency = number of inspections conducted annually.

Water Age Management

- Pipeline maximum water age = days resident in the system distribution mains (calculated by a calibrated hydraulic network model).
- Storage facility maximum water age depends on operating conditions (calculated by a calibrated hydraulic network model).

Water Loss Control

- Infrastructure condition factor—see AWWA Manual M36, *Water Audits and Loss Control Programs*, for details.
- Infrastructure leakage index—see AWWA Manual M36 for details.
- Annual real losses—see AWWA Manual M36 for details.

Pipeline Rehabilitation and Replacement

Pipeline rehabilitation and replacement (R&R) has a significant effect on water distribution system infrastructure condition and performance. By targeting older or problem pipelines for renovation, utilities can reduce leaks and breaks, improve water quality, and enhance water delivery reliability. Management planning for major pipeline renovation programs is described in this section.

Pipeline R&R Plan

Infrastructure renovation (replacing and rehabilitating pipelines) is an expensive undertaking. Utilities usually cannot afford to replace aging system components at a rate they would prefer and must make choices regarding the amount of system improvements they can perform annually. A pipeline R&R plan should be developed to provide priorities so that the most needed improvements can be made while others must be delayed. The primary components of this plan are preparing a pipeline inventory, conducting a performance assessment and tracking, and determining priorities.

Pipeline Inventory An inventory of pipelines with physical and location attributes is an essential element of a pipeline R&R plan. Data in the inventory should minimally include pipe size, material, class, lining, and year installed. Location data should include street name, municipality, and pressure zone. The effective management of physical asset and location information for pipelines can enable a utility to locate and repair mains and plan for rehabilitation or replacement.

Performance Assessment and Tracking Performance information may include customer complaints, breaks and leaks, and water quality, flow, and pressure. The frequency and type of main breaks and leaks are often the most important performance information. Several methods are available to assess the pipe condition that may lead to failure. Some of the more common methods are soil corrosiveness testing, pipe coupon testing, and acoustic leak detection. Video inspection of pipelines is also common practice for large mains.

Planning and Prioritization A pipeline renewal program should be coordinated with the utility capital improvement plan. The program must consider the appropriate annual funding and select the specific main segments to be renewed. Because of financial constraints, most utilities must limit the number of renewal projects to only those that have the highest priority.

AWWA Manual M28, *Rehabilitation of Water Mains*, describes several approaches for prioritizing renewal of mains, and some models are commercially available. These approaches include scoring methods (e.g., point systems), economic analysis (like break-even, cost–benefit), failure probability and regression analyses, and mechanistic models to predict pipe failure based on loading and condition.

Operational Considerations

Utility management is usually responsible for developing a pipeline R&R program and for the decisions to implement the plan due to the cost and impact of these programs. System operators, however, have an important role. Operators must often acquire the system inventory data needed for constructing performance measures and setting renewal priorities. As projects develop, they may present situations involving water supply interruptions or unusual operating conditions. System operators should plan for these adjustments to minimize adverse customer impacts.

Employee Management

Policies and Regulations

As noted in Chapter 1, there are many regulations affecting water quality and operations associated with water and wastewater utility systems. In addition to the capital assets of the utility system, and the maintenance of those capital assets, discussed elsewhere in this text, a water utility has another significant asset—its employees. Sustainability of the utility involves maintaining its infrastructure and its personnel assets (Ralston and Ginley, 2010).

Just as the capital assets must be protected and maintained, the personnel assets must be protected and nurtured. This involves acquisition, training, and development so employees contribute positively to the utility mission. Personnel have knowledge of the utility, an attribute that can be capitalized on by developing an appropriate work order tracking system that transfers information discovered during breaks and other maintenance activities to paper, and ultimately some form of electronic analysis.

Just as there are many regulations in the operation of the utility assets, there are even more associated with the human resource aspects of utilities. Some are rules and regulations, while others are policies. Enforcement differs depending on the type. The ability to change, alter, or ignore them varies as well. Regulations have legal consequences, while policy implications do not. Most personnel rules are regulatory in nature. However, purchasing is usually associated with policy implications.

A **policy** is a guideline that employees are expected to follow. Decisions are made on policies that are generally defined by a governing board or executive management. The goal is to ensure consistency in decisions relating to a given situation (e.g., the utility always turns the water off if people fail to pay their bill in a timely fashion, as opposed to turning off only certain people's water, which would be unfair). An inconsistent policy could cause people to question the utility's policy-making ability.

Policies should provide some capacity for interpretation, however, which is why they are deemed to be guidelines as opposed to rules. Policies allow the employees and, at times, public to understand what is expected and what reasonable implementation is to occur. For example, most utilities have a purchasing policy that requires a certain number of quotes or bids to ensure the utility gets materials and work performed at reasonable prices. However, in the event of an emergency, the utility does not have to follow those rules. It could address the situation without following the policy, especially when life, safety, and welfare of the public might be concerned.

Policies Affecting Supervisors

There are three important policies a supervisor should be aware of. The first is budgeting policy. The supervisor is likely the person who prepares the initial budget for the areas under his or her control. Money must be available to pay for the work that is planned; thus a sufficient amount of money needs to exist for this purpose. The supervisor knows the work better than anyone else in the organization and may need to explain what is needed. During the year, the supervisor is responsible for the judicious allocation of funds in order to accomplish the tasks at hand. This includes both labor and materials.

Materials lead to the second policy; there needs to be a written policy for how to purchase materials, tools, and so forth. All are routine items that employees will require to accomplish their work. If there are no purchasing rules, the potential exists for the organization to pay more for materials than it would otherwise need to. In the public sector, there are usually state laws that outline how municipalities are to purchase tools, equipment, and other services. Even for private-sector utilities, there should be a program to ensure that multiple sources are contacted and that the lowest prices are accepted as long as they meet the requirements to be responsive and responsible. These terms mean that the bidder has the equipment or services to adequately provide the work or that the materials meet certain specifications or qualify within the utility's guidelines; that is, the bidder can provide the appropriate equipment and can complete the service by the specified date.

policy
A guideline that employees are expected to follow.

The supervisor also needs to understand personnel. There is a protocol for hiring personnel, monitoring their efforts, and supervising them. Also, there may be union rules regarding hiring or the hours that personnel may be scheduled to work. Most labor rules are based on laws and statutes approved by elected bodies and must be complied with. The supervisor should keep in mind that there may be regulations associated with the work. It is the responsibility of the supervisor to ensure compliance with the regulations. Failure to comply with regulations falls on the supervisor, even if it is the employees who are violating them. The supervisor needs to take action to ensure policy compliance. Ultimately, it is the supervisor's responsibility to ensure employees are trained and advised of relevant situations. Discussion of the basic employment laws follows.

Employment Laws

Employers and employees need to be aware of regulations associated with personnel. Employers are required to post notices to all employees advising them of their rights under the law, including Equal Employment Opportunity (EEO) regulations. Such notices must be accessible to the employees regardless of disabilities. There are several acts of particular importance from a federal perspective, including the Civil Rights Act of 1964, the Equal Pay Act of 1963, the Age Discrimination and Employment Act of 1967, the Americans With Disabilities Act of 1990, the Rehabilitation Act of 1973, and the Civil Rights Act of 1991.

Workplace Safety

There are a variety of rules for employers regarding the safety of workers and the workplace. Before these laws, unsafe working conditions often resulted in employee injury. Productivity of the employees and businesses were adversely affected, and employers often bore the additional cost of rehabilitating or compensating the injured employee. In some cases, employees were killed or permanently disabled as a result of workplace activities.

The Occupational Safety and Health Administration (OSHA) was created to help minimize the number of workplace accidents and injuries and to require employers to provide safe conditions and appropriate training. It is recognized that not all workplaces can be made completely safe. An example would be a water plant or a construction site. By their nature, accidents can occur and working with equipment can lead to accidents.

Because operations personnel come into contact with construction, road work, chemicals, and sewage, these workers should be cognizant of the need for proper use of hard hats, safety vests, eye protection, hearing protection, safety shoes, and safety gloves. Respiration devices are required for entry into many confined spaces. Confined spaces require entry training, multiple-person crews, and airspace testing. Ongoing training is required to reduce workers' compensation claims, lost time, and legal liabilities.

Also required are a variety of rules involving chemical usage and handling, hazardous material safety data sheets (MSDSs), regular training of staff in the proper use of chemicals, chemical cleanup procedures, and emergency leak response plans. Chlorine is often the most dangerous chemical used by utilities, but safety training is often readily available through chlorine dealers.

One person should be assigned safety responsibilities in larger utility systems. It is best if this person reports to the utility manager as opposed to a field supervisor for two reasons: safety is elevated in stature when management is involved, and there is less potential for conflict between field activities and safety procedures.

It is vital to minimize the frequency of adverse incidents by ensuring that employees are not subjected to undue hazards and that appropriate training is provided. Failure to comply with proper procedures, on the part of either the employer or the employee, may bring OSHA into action. Employees have a right to notify the employer or OSHA about workplace hazards. The reporting employee's name can be kept confidential. Employees have the right to request that OSHA provide an inspection of a workplace if it is believed that unsafe or unhealthy conditions exist. In addition, employees have the right to file a complaint with OSHA if, as a result of a complaint, retaliation or discrimination has occurred. Employees have the right to see any OSHA citations that may be issued against the employer for unsafe conditions. In part, this is permitted to allow the employee to determine if there is a recurring pattern or if an accident is an isolated incident. OSHA treats these situations differently. If a citation has been issued, the employer is required to post the citations near the alleged violation site. The employer is required to correct workplace hazards as identified and in compliance with the date on the citation, and it must certify that the corrections have been completed. If the employees have been exposed to harmful conditions or toxic substances during the period of the violation(s), then they have the right to have copies of this information provided to them and to their physicians.

Supervision of Operations

Defining Supervision

A supervisor is responsible for the efficient direction and use of human and material resources in such a manner as to effectively produce goods and services to meet customer needs. Some terms associated with effective supervision are *optimal*, *efficient*, and *low cost*, which further increase the responsibilities of and expectations for the supervisor.

Supervision could be defined as using available resources, human capital, materials, equipment, and funding to meet the organizational goals. The proper allocation of these resources—i.e., knowing when to use them and when not to—is one facet of successful supervision. The supervisor is responsible for developing goals for the employees under his or her control, in keeping with the goals of the organization, and for directing and allocating the necessary resources to permit the employees to achieve those goals. If the efforts do not appear to be working toward achieving organizational goals, then the supervisor must make necessary changes to improve the likelihood that the goals are met.

Personnel Management

Without supervision, basic direction, and an orientation toward achieving organizational goals, chaos would result. Everyone would do what they felt was most beneficial to them, which would not provide any cohesive effort or include any organizational goals. Frederick Taylor, the "Father of Scientific Supervision," tried to convince organizations to achieve the following four objectives (Taylor, 1911; FWPCOA, 1997):

1. Increase physical output.
2. Decrease production costs.
3. Lessen employee fatigue.
4. Increase wages.

At the time when Taylor made these recommendations, the average wage was minimal and employees worked many hours to acquire those wages. Modern manufacturing processes were not invented until the 20th century when Henry Ford invented the assembly line. There was limited mechanical power, so the employees did everything by hand. Most employees likely believed there was little or no hope of improving their lot in life. As a result, there were no incentives to work harder, because working harder simply meant doing more work, a disincentive to employees. Automation improvements in manufacturing processes improved profits, productivity, and efficiency, and provided less-expensive products that spurred other economic activity. Thus, Taylor saw the innovations in the workplace as an opportunity to improve conditions for employees while motivating them to achieve still greater productivity.

Part of supervision is getting the most out of people, which means letting them develop to their fullest potential. The advent of new ideas implemented to improve manufacturing productivity created both the need to maintain equipment to ensure continued productivity and the need to improve labor skills. The way to improve labor skills was to improve education through training and public schools, and to invest in technology that would minimize adverse work conditions and lower the potential for injury to employees. It was realized that trained employees were assets (human capital) just as much as equipment, buildings, and products. As a result, the investments in these assets needed to be protected because training new employees often required hours of downtime. Having skills also became more of an issue in hiring employees because profits were tied to efficiency.

To understand the change that occurred during this period, one needs to look at the specifics of the evolution of technology and productivity. The need for effective training increases as technology advances. Technology is one of the major factors that has changed the workplace, starting with electricity and moving to assembly lines, transportation changes, computers, and robotics. At each point, the intent of the change was to improve productivity while reducing human error and interface. One could view these improvements as a path to cutting jobs, but the opposite is true. What technology does is reduce the demand for unskilled-labor jobs, which are much more efficiently performed by machines.

Ditchdigging is an example in the water service of how technology has improved productivity while also improving work conditions for people. A four-man work crew is common for water distribution pipeline crews. How productive would these people be if they had to lay all piping by hand? How much pipe could they lay in a day? Working diligently, it is doubtful that they would lay more than 50 ft (80 m) of pipe, let alone backfill it properly. They can lay the pipe—the easy part—but it is the digging that is the problem. So backhoes were developed through advances in technology. The backhoe eliminates a substantial number of unskilled ditchdigging jobs, as it, with the appropriate operator, can dig as much as 1,000 ft (1,600 m) of trench in a day, and backfill it. To achieve similar results, it would take a crew of 100 or more people. Yet, the backhoe cannot put the pipe together. Putting the pipe together requires manpower to chain it for lowering and placement. The backhoe can help by lowering the pipe, and perhaps pushing it together, but the backhoe cannot guide the pipe. The number of pipe layers remains the same, but technology has put a lot of ditchdiggers out of work. In place of these jobs, people are required to operate the technology and service/repair it, which are jobs that did not exist before the technology was invented. These are higher-paying jobs, for which training is needed.

Technology and training help utility personnel do their jobs more effectively and, in most cases, more efficiently. When training, technology, and work conditions improve, higher effectiveness will follow. More highly skilled workers with proper training can do their work more efficiently, which improves morale. Greater efficiency benefits the organization as a whole because unforeseen costs should decrease.

The fact that there is less physical labor involved in the field means more inside work is needed to maintain complex technology (more training), which improves work environments. Less time is required outside in the weather, while more time is spent inside in the air conditioning using more SCADA and more computers. Better working conditions and higher-training-level requirements will attract different types of workers, who will seek higher pay and better benefits and may accomplish more as a result of new technology. Higher output means more can be paid for skilled workers. Overall work conditions improve as a result.

Organizational health is an important driver, especially for good employees. As Collins notes in *Good to Great: Why Some Companies Make the Leap . . . and Others Don't*, great organizations strive to place the best-qualified people in the proper positions (Collins, 2001). The concept is that people who have the skills and drive to do their work will do it well and be happy. They will work as a team, demonstrating respect for one another, while solving organizational issues. The key is making the organization, not individuals, great. Success for the organization will ultimately mean success for the individuals.

Organizations that just "fill positions" typically are more hostile. Hostile environments are not conducive to effective or efficient operation. In addition, organizations that attempt to resolve discourse through replacement with "yes people" fail to adapt to changes in the field, which impinges on their ability to remain effective or efficient. "Good and great" supervisors are looking for people who strive to be great, who have the requisite skills, and who can provide perspective on their areas of responsibility (Collins, 2001). Having differing opinions is good, assuming some consensus can be reached, as opposed to infighting. Turning these differences of opinion into a constructive, rather than destructive, force is where the supervisor rises to achieve effectiveness in the organization, or fails personally and organizationally. It is the latter issue—their responsibility is to the organization—that supervisors often do not understand.

Great supervisors surround themselves with great employees (Collins, 2001). Great supervisors advocate hiring people who can do more than the job they are being hired for. Great supervisors often look for employees with more skills and potential than they themselves have. The reason is that such people will be successful. They will ultimately leave the position, but they likely move upward in the organization.

Factors Affecting Supervision

There are a variety of factors that impact the supervisor's ability to accomplish his or her goals effectively. One is economics. If the economy is poor, pay increases or bonuses are difficult to provide. As a result, economic turmoil, recession, and depression will adversely affect the supervisor–employee relationship. Employees have little option for movement, so perceive little bargaining power with employers. Societal adjustments are also a confounding factor. Younger employees have different work ethics and different incentives than older employees. They value free time and less rigid work environments. Older employees accept rigid work schedules and may be more motivated by pay and security.

Legal, political, and regulatory issues are often related to one another. In the early 20th century, major inroads were made to improve working conditions, especially for textile workers. Child labor was eliminated and education was encouraged for all children. Today the Fair Labor Standards Act limits work hours and requires overtime pay beyond a 40-hour workweek. Equal-opportunity laws prohibit discrimination in hiring employees, and OSHA requires employers to provide safe work environments and train employees on workplace hazards. Supervisors must understand the implication of all of these types of laws and policies to the organization and to themselves personally. Supervisors who do not comply with these rules can be held personally liable.

Given the responsibilities and restrictions, and based on the reasons why a person would want to be a supervisor, problems can occur when operations personnel are promoted from within. Line personnel often have relationships with one another, often close ones. This is normally useful in the field, but it is not appropriate between supervisors and those under their supervision, and remains a common problem for those who are promoted from within. One cannot maintain the same relationships with one's old "buddies" and supervise them effectively. A common reason for failure among supervisors promoted from within organizations, and a major reason that management does not often do it, is that operators who are promoted from line positions to supervisors cannot overcome their roots.

A part of supervision is the ability to adjust to a dynamic work environment, something that line employees are less comfortable with. A criticism often levied on people who have been in supervisory positions with utilities for many years is that they still do things the "old way" and have not updated their skill set. Dynamic supervision requires that the supervisor understand that different circumstances and different people will require different approaches. Successful supervisors do not take a "one-size-fits-all" approach to management. They seek to continuously improve themselves, to incentivize staff, and to allocate the people and resources at their disposal to meet organizational goals.

Supervisory Functions

Not only does a supervisor need to have a variety of skills that must be judiciously implemented with the resources available, but he or she must also understand the supervisory functions. These functions include planning, budgeting, organizing, staffing, directing, encouraging communication, controlling, managing time, and decision making. Optimizing the performance of the organization and accomplishing organizational goals depend of the successful implementation of these functions (FWPCOA, 1997).

Planning and Budgeting

From a planning perspective, it is the responsibility of the supervisor to plan the work that is to be accomplished each day and over short- and long-term horizons. It is necessary to have a series of long-term goals for the organization so that the supervisor can make decisions and plan work so that all components necessary to achieve these goals occur at appropriate times. Planning requires not only understanding the goals, but understanding what is necessary to accomplish them in terms of proper utilization of employees, equipment, materials, and other resources.

The supervisor's success in planning is indicated by whether the goals are met and whether there is a minimal amount of wasted time, effort, or other resources. For example, you can have a great crew—highly motivated, efficient, and well equipped—but if they arrive to a worksite only to find that someone forgot to order the necessary parts, then this shows a lack of supervisory planning.

The planning horizons are short and long term. In many cases, decisions made in the short term affect the long-term ability to accomplish certain tasks or to complete certain types of work. Short-term goals may involve days, weeks, or months. For a line supervisor, many of the decisions are day-to-day implementation and utilization of resources. Longer-term planning horizons may be goals to complete by the end of the month or by the end of the fiscal year, whereas short-term goals might be completing preventive maintenance work, performing monthly operating report gathering, and daily sampling. Longer-term actions might include replacement of pumps or the shutdown and repair of certain treatment plant components. Long-term actions might be oriented toward plant expansions or new water sources that might be as many as 20–50 years out, but these are primarily management issues, not line supervisor responsibilities.

A fourth area of planning involves understanding what emergencies or conditions might occur, what breakdowns could occur, and where pipes and equipment might fail. Anticipating such events will help with creating contingency plans, which are part of the supervisor's responsibility. The line supervisors are typically responsible for implementing the emergency plans and taking actions associated with vulnerable infrastructure. A proper plan, whether it is documented on paper or created on a day-to-day basis within the supervisor's head, requires that goals be established (e.g., what is to be accomplished today, tomorrow, this week, this month, and this year) with the understanding of the current situation.

Once the goals are established, it becomes clear what tasks need to occur in order to get from the current situation to the hoped-for future condition. Understanding barriers or issues that might prevent accomplishment of those goals in the future will become more obvious as those completing the work continue to plan and communicate with one another. To facilitate communication among workers, the supervisor should write plans down versus, for example, planning in one's head.

Once the goals, present condition, and barriers to accomplishing the future are understood, an action plan can be developed whereby resources are allocated, materials ordered, and staff and equipment provided to accomplish the task. A budget will be developed so that it can be determined what the various tasks may cost.

Looking at an example, operators take water samples on a daily basis to monitor the treatment process, measure chlorine residuals, and take biological samples. They also monitor equipment. Thus, on a daily basis, there should be a plan laid out so that everyone knows when these tasks are to occur and what results are expected. In a treatment plant, it might be that a jar test is conducted on an hourly basis to determine whether the coagulation process is functioning properly. Or, on an hourly basis, a chlorine residual test might be taken. Such tasks can be written out on a schedule so that everyone knows that every hour this test must be completed (and appropriate employees assigned to do the work). In certain months, the operating data and daily operating data must be recorded; hence, once a sample is taken, data must be entered into some database to indicate that the work was actually completed. This should also be scheduled.

In doing an annual budget, it is clear how many tests will occur and how much effort it actually takes to accomplish them, so those resources can be allocated. Therefore, the short-term goal of taking samples hourly, recording the results, and determining what the long-term set of results will look like provides data so budget costs can be assigned.

A repair, such as to a pump, is an action that would occur rarely. The supervisor would have to determine how much downtime there might be, what resources are needed, and what contingencies could occur. In this case, contingencies could include taking longer than expected to get the motor rewound, bolts that do not allow the pump to be moved as efficiently as anticipated, bolts that break off, and/or a pump that actually cannot be repaired. All are potential issues that would keep the staff from accomplishing their goal in a timely fashion. Necessary resources that might need to be scheduled include equipment (e.g., a crane), an extra person or two to offer support as needed, and electricians or others to connect or disconnect the pump. The supervisor can then figure out what the cost of these factors would be and schedule everything accordingly.

Once the plan is created, the action is taken and results are verified. If the supervisor does not verify that the results actually occur, as anticipated with regard to resources and budgets, then the planning process never has a balance to determine the effectiveness of actions. Work order tracking and maintenance management systems have significant budgetary value in determining the frequency and cost of repairs.

A lot of work on the part of the supervisor goes into planning and it may be one of the most important aspects of the supervisor's job. Planning work and budgets and making sure materials are on hand are activities that will take some thought and time to accomplish properly. Meetings, even if very brief, to discuss what is expected and how it is to occur are generally required. And, of course, supervision is required to ensure that the work is actually done.

Organizing and Staffing

Organizing for the work involves defining the following: the task to accomplish, the human resources required, the skill set required, materials, and the work environment where it will be accomplished. Most of these elements are obvious, but the work environment may be more difficult to define. The work environment may be a normal site during the day or it may be scheduled at a specific time and place. For example, a pump may need to be changed at midnight when flows are low because it needs to remain in service during the day. If the utility is doing work at night, extra lighting may be required, people may need to be brought in after hours, and so forth. Therefore the environment affects all those decisions in regard to resources required.

The supervisor must have authority from management. If the staff does not believe that the supervisor has management's authority to direct the work, then the work may not get done efficiently. In this case, authority means delegation of responsibility from management to the supervisor, which means there is an expectation of certain work being accomplished. The supervisor should delegate the responsibility for task completion to the line personnel. The supervisor should not have to watch to make sure every task gets accomplished; he or she delegates that to the staff and holds them responsible for a task's completion. For example, if operators fail to do the hourly sampling tests mentioned earlier, then they need to be questioned as to why and perhaps disciplined for failure to accomplish this task. Ultimately, this is a team-building exercise in which the team is only as effective as its weakest member.

Supervisors want to build an effective team, which requires that the team understand the work they will do and what is expected at the conclusion. Productive, happy employees are enthusiastic about what they are going to accomplish and are typically more productive and efficient in accomplishing tasks than those who are not. This fosters teamwork because everybody appears to be pulling their own weight and the employees believe they can count on one another. Teamwork requires that the supervisor not micromanage. Tasks are delegated so employees feel some empowerment and have the ability to make choices and decisions that affect what they are doing. In many cases, supervisors, especially young ones, do not understand that many decisions have results that are inconsequential to the accomplishment of the organizational goals but may have a very significant impact on the person who is actually doing the task. Choice of equipment is a good example. A group of employees may be much more confident in getting work done using one piece of equipment than another. As a result, if it is critical to get this work done, the choice of equipment used may be of minimal significance to the supervisor but important to those employees.

Directing

Direction is defined as the issuance of orders, instructions, assignments, and the necessary guidance and oversight/follow-up. Direction is useless without the following:

- Planning
- Evaluation of talents
- Acquisition of tools and materials

To use a football team as an example of the process, it is accepted that the goal is to score touchdowns and win the game. The quarterback on the field is directing the action, but he has a supervisor, the coach, who is directing him (calling plays and sending in the personnel to enact those plays). The coach instructs all the players and tells them their responsibilities. If everyone accomplishes their responsibilities, they will score touchdowns. The more effective they are, the sooner this will occur. Once on the field, though, the quarterback assumes the supervision of what happens. He may change the play; he may change direction; he may change receivers to whom the ball is thrown based on the conditions on the field. But the team has to be on board with these decisions. If they are not, the quarterback gets sacked or throws an interception and progress is stopped.

As in sports, groups that work as a team are more effective than they would be otherwise. When everyone understands and performs their role, the team succeeds. And, as in sports, those who cannot comply are usually cut. If they are winning, everybody is happy. Productivity is winning.

Supervision requires a lot of planning, effective use and evaluation of talent, and acquisition of tools and materials, all of which take a substantial amount of time and thought. Some employees do not understand how much time and effort are required of the supervisor; likewise, some supervisors, when they take the job, do not understand these requirements. If insufficient time is spent on planning, the organization will suffer.

Direction involves assigning responsibility for the work to be done. The supervisor must make it clear which team members are responsible for which tasks and must ensure that everyone understands what is to be accomplished. There must be a timetable for the work's completion, possibly with intervals for each task. The team must accept the job responsibilities, which means they understand the tasks at hand and will perform them (Diamond, 2007). If there are employees who refuse to accept responsibility, then they ultimately are probably not beneficial to the organization. But most employees, if they believe that their opinions matter and their skills are used, will accept responsibility for their piece of the operation.

Directions must be clear and reasonable. For example, "You must lay a 10,000-ft water line in three days," is clear but not reasonable. Unreasonable direction will not be conducive to the organizational goals or team building (Diamond, 2007). In addition, assigning work that has no relationship to organizational goals will cause the supervisor to lose the respect of his or her crews, and the work is unlikely to be accomplished (employees realize that failure to accomplish goals falls on the supervisor). It is the supervisor's responsibility that the work expected to be done is understood.

Ensuring Efficient Communications

Miscommunication is the most common problem in the workplace. The common response to work not being done as the supervisor intended is to have lengthy discussions with the employee or employees responsible. Unfortunately, it may be that the supervisor failed to convey the information properly, not that the employees failed to do the job as intended. Failure to communicate only leads to hard feelings on the part of the employees and frustration on the part of the supervisor. Hence, it is important up front to make sure everyone understands what is expected.

Teamwork skills are required in any organization. Individuals who go off on their own are rarely successful, although they may be successful in accomplishing personal goals. In part, good interpersonal relationships build an opportunity for positive progress and goals. For example, sports teams with divisive locker rooms are often not successful (i.e., not a winning team). The issue is directly akin to that in virtually every other workplace. It is necessary that appropriate expertise and resources exist within the team and that team members understand who is responsible for which resources and where to go for the necessary expertise. The goals for the team must be clearly defined and obtainable. Unrealistic goals will hurt teamwork because the team will believe it is doomed to fail. There is a need for management at all levels to be supportive of the goals of the team, provide leadership, and communicate to the team (Diamond, 2007). The expectation is that the team will be able to communicate among themselves. This does not mean that they need to be friends, but that they are able to work together to accomplish the greater good. If each employee does his or her share, the combined efforts are of greater value than if the same individual efforts occur independently. Typically, teams that work together will show a limited amount of interpersonal conflict.

Controlling

Control is another of the steps required to ensure that the work gets accomplished and that progress toward the goals of the organization is ongoing. Control is mostly about standards. Questions such as the following might arise after an activity:

- Was the time to accomplish the work reasonable? Was it too long or too short?
- Did the team work effectively?
- Did the personnel perform satisfactorily?
- Did they all do their jobs as they were expected to?
- Was the cost reasonable or did they end up spending a lot more time and effort than was needed, or was additional equipment required that was not anticipated?
- Was the job done satisfactorily or were there issues?

To determine whether these goals are met, the supervisor needs some form of feedback, which is why tracking time, money, and resources on a work order is required as a part of the process. Once the work is complete, an evaluation can be made to see if the plan and the results coincide. It may be that certain things occurred that were not planned, the budget was overspent, or the time required exceeded expectations. The supervisor needs to be reasonable in his or her evaluation. His or her estimates of time may not have been correct or estimates of budgets may not have been correct. The process is to plan it, do it, determine whether it met the plan, and, if it did not, improve outcomes the next time or for the next portion of the plan.

Managing Time and Making Decisions

Time management is one of the components that must be addressed to properly allocate resources. Often this involves delegation. Many supervisors have difficulty delegating, whether because they fail to do so often enough or because they do so too liberally. Delegation involves assigning responsibility to someone else to accomplish a task on time, on budget, and with relatively quick decisions. There are other decisions that are delegated as a part of a task, which may include spending money, ordering materials, or determining how work is to progress. Expectations must be set.

Time management involves deciding simple things quickly; deciding who should attend what meetings; learning to avoid taking on issues that consume large amounts of time but are not within the core mission; moving forward immediately on tasks, including doing the difficult parts first so that the easy parts come later; and traveling lightly. Time management also involves avoiding tasks such as creating useless e-mails or memoranda as opposed to addressing or solving a problem. Avoiding unimportant tasks, looking forward to and anticipating upcoming tasks, and minimizing meeting and telephone time are important. Managers also need to ensure that their staff does not procrastinate in completing work or making decisions.

To make decisions, it is necessary to determine what the situation actually is. In many cases, a situation is poorly understood due to lack of data, experience, or time to respond. If the situation is not understood, decisions made will usually be in error. If the situation is understood, data can be generated and analyzed. The analysis should provide some direction about the choices that can be made. For example, in the treatment plant, a funny noise is heard in the pump room. It could be a variety of things, including motors, pumps, or valves. It is necessary to determine exactly what the situation is, as opposed to immediately assuming there is a bad bearing in a pump somewhere and starting to take pumps apart. If it is a clattering check valve, then that is a very different repair than replacing pump bearings. Once the action is taken, the choice needs to be reviewed and, if further action is required, that action can be taken. Making decisions means determining a direction from multiple options, and it is the supervisor's responsibility to choose the best option.

Budgets

Budgets were discussed previously in conjunction with the supervisor's planning responsibilities. Budgets are discussed here in a broader sense as a necessary part of utility operations. All utilities should be set up as an enterprise fund in order to allow them to pay their own way. Budget preparation is usually done by line managers and reviewed by supervisors, utility management, a budget/finance person, and the management of the utility or city. The budget is a planning tool and an organizational tool. It should not be set in stone. Many people have difficulty understanding that the budget is simply a plan, and, especially with an enterprise fund, the plan can be adjusted upward or downward to accommodate unforeseen conditions.

Table 12-2 is an example of a budget that includes capital within the budget. It is important to note that for many of the items, a simple increase in percentage each year is not appropriate. Past costs may be indicative to some extent of future costs, but at the same time, note that certain line items in this budget have varied significantly, mostly in the maintenance area. If a pump goes down, it needs to be repaired, but repairing multiple pumps in a given year may not be planned for. As a result, having retained earnings is an important part of the utility operation. Retained earnings is the amount of money that is set aside as "surplus" money that is not expected to be spent as a part of the current budget. Retained earnings are important from an audit perspective, from the perception of the financial community, and from an operating perspective. It's clear from a political level that governments that do not have retained earnings have difficulty responding to the needs of their constituency during times when the economy deteriorates.

Many depression-era and modern-day economists note that when the economy takes a downturn, there is a need to continue spending levels to keep people employed. The problem is that most local governments and utilities, when overcollecting revenues, tend to reduce their tax levels and reduce their fee charges to balance the budget instead of adding to retained earnings. The result is that when the economy becomes difficult, they need to raise taxes and raise fees precisely when their constituency has little ability to pay for it or change levels of service. Additionally, when the economy is in a down cycle, there is a tendency for construction projects, the major cost of capital for utilities, to decrease. Having significant retained earnings allows the utility to operate consistently and perhaps even act as a mechanism for increasing jobs in the construction industry through a likely backlog of projects. As a result, the utility may find itself saving a significant amount of money by constructing needed infrastructure during poor economic conditions.

Table 12-3 is a project budget. The project budget is used to look at initial costs and life cycle costs. Reviewing these costs is necessary in order to evaluate different options. The budget process is what links the operation, management,

Table 12-2 Example budget that includes capital within the budget

Account Number	Account Description	2007 Actuals	2008 Last Year's Actuals	2009 Original Budget	2009 YTD 5/30/09 Actual
403-3801-538.31-10	Salaries & Benefits	$388,302	$425,013	$471,512	
	Professional Services	14,876	5,879	26,271	7,499
403-3801-538.31-30	Consultant Engineers	20,218	33,109		
403-3801-538.32-10	Accounting & Auditing	1,132	867	960	917
403-3801-538.34-10	Contractual Services	3,661	1,190	1,056	740
403-3801-538.40-10	Training & Per Diem	1,306	2,090	1,988	1,188
403-3801-538.40-30	Expense Account	601	603	600	420
403-3801-538.41-10	Telephone	387	636	780	375
403-3801-538.43-20	Electricity	8,076	7,133	9.966	3,967
403-3801-538.44-10	Equipment Rentals	274	255	255	170
403-3801-538.44-20	Trailer Rentals	900	900	900	675
403-3801-538.44-30	Misc. Rentals	977	992	1,010	1,042
403-3801-538.45-10	Insurance Coverage	6,271	6,139	5,142	4,982
403-3801-538.46-10	Equipment Maintenance	15,549	8,113	8,000	14,500
403-3801-538.46-30	Vehicle Maintenance	13,111	15,747	12,000	8,182
403-3801-538.46-50	Grounds Maintenance	21,569	18,573	18,000	10,608
403-3801-538.46-60	Utility Maintenance	14,498	10,040	52,000	1,350
403-3801-538.49-30	Permits & License Fees	4,149	5,543	4,984	13,527
403-3801-538.51-10	Office Supplies	145	150	150	0
403-3801-538.52-10	Gasoline	8,782	11,125	15,848	4,859
403-3801-538.52-20	Misc. Supplies	1,037	–24,034	7,000	2,648
403-3801-538.52-50	Uniforms	540	559	750	700
403-3801-538.52-60	Building Supplies	275	0	0	0
403-3801-538.62-10	Buildings	0	0	162,488	18,244
403-3801-538.64-00	Machines & Equipment	46,919	(1,843.00)	0	0
403-3801-538.64-20	Vehicles	23,219	0	0	0
403-3801-538.64-30	Capital	23,700	1,843	3,500	0
403-3801-538.71-10	Principal	0	0	16,446	12,366
403-3801-538.72-10	Interest	4,029	5,694	2,666	2,812
403-3801-538.91-10	Transfer to General Fund for purchases and services	0	0	30,000	20,000
403-3801-538.99-10	Contingency	0	26,030	181,195	0
		———	———	———	———
	TOTAL OPERATING	$624,503	$562,346	$1,035,467	$131,771

2009 CMPR Anticipated?	2010 Prop Budget	2011 Prop Budget	2012 Prop Budget	2013 Prop Budget	2014 Prop Budget
$427,077	$458,486	$541,410	$568,481	$596,905	$626,750
11,300	11,300	6,000	6,000	6,000	6,000
	50,000	35,000	35,000	35,000	35,000
0	1,050	1,082	1,114	1,147	1,182
1,040	20,840	21,465	22,109	22,772	23,456
2,000	2,000	2,060	2,122	2,185	2,251
600	600	618	637	656	675
648	648	667	687	708	729
9,966	7,409	7,631	7,860	8,096	8,339
255	255	263	271	279	287
900	900	927	955	983	1,013
1,094	1,094	1,127	1,161	1,195	1,231
5,553	5,855	6,031	6,212	6,398	6,590
10,000	10,000	10,300	10,609	10,927	11,255
12,000	12,000	12,360	12,731	13,113	13,506
18,284	18,248	18,795	19,359	19,940	20,538
60,000	60,000	61,800	63,654	65,564	67,531
20,000	20,000	20,600	21,218	21,855	22,510
150	150	155	159	164	169
11,662	11,663	12,013	12,373	12,744	13,127
7,000	7,000	7,210	7,426	7,649	7,879
750	750	773	796	820	844
0	0	0	0	0	0
0	0	0	0	0	0
0	0	0	0	0	0
0	0	0	0	0	0
0	3,000,000	600,000	125,000	230,000	145,000
17,183	23,507	227,763	267,876	267,876	267,876
1,930	1,930	incl above	incl above	incl above	incl above
0	45,000	47,250	49,613	52,093	54,698
0	50,000	0	0	0	0
———	———	———	———	———	———
$619,392	$3,820,685	$1,643,299	$1,243,421	$1,385,069	$1,338,435

Table 12-3 A project budget

Cost Item	2009	2010	2011
Planning	$25,000		
Phase I—Sewer investigation and repair (G7 Program)	$312,000		
Phase I follow-up		$75,000	
Letters to homeowners		$5,000	
Identification of areas to televise		$25,000	
Phase 2—Infiltration repair (televising and lining)			$975,000
Phase 3—Point repairs			$150,000
Total cost of project			$1,567,000
Annual cost savings $575,000			
Present worth of cost savings (20 years)	$8,554,548		
Debt cost for program (4.5% for 20 years)	$120,465		

line item
An expected cost within an expense category.

annual audit
A financial review that comes after the budget year has been completed in which an external accounting group evaluates whether the revenues and expenditures were appropriately categorized, money was maintained in the appropriate accounts, and revenues and expenses were fully accounted for and appropriately spent. Also known as *comprehensive annual financial report* or *CAFR*.

and finance aspects of the utility. As the interactions of these three management concepts basically parallel one another, capital management similarly is interwoven among these three groups.

While the budget is a plan, it is also a mechanism to plan expenditures and allocate them among different expectations of expense categories, commonly referred to as **line items** (see Table 12-4). An allocated budget for operations and maintenance will permit the utility to evaluate its competitiveness and determine areas where costs might be saved. For example, if energy costs are significantly higher for the utility than one might expect, then the utility can start looking for areas where it may have pumps, motors, drives, aerators, or other equipment that is not very energy efficient and can be replaced with more efficient equipment, thereby saving money. In some cases, power companies and the federal government may offer grant programs to help utilities lower their energy costs.

The outcome of the budget process indicates the necessary revenue needs for the utility and provides some indication of whether the maintenance costs of the system are increasing, perhaps because of deteriorating conditions of the assets.

The budget will be adopted by a resolution of the governing board and can be modified only through amendments approved by the governing board. However, the mechanism for approving the budget should not make each line item a not-to-exceed amount that needs to be modified through board action. That way operations managers and personnel can move costs between line items, many of which are small (see Table 12-4), without having to go through the process of gaining approval for moving monies within a given budget.

The **annual audit** (comprehensive annual financial report, or CAFR) comes after the budget year has been completed. The intent of the audit is to have an external accounting group evaluate whether the revenues and expenditures were appropriately categorized, money was maintained in the appropriate accounts, and revenues and expenses were fully accounted for and appropriately spent. The audit will contain useful information about the revenues and expenditures for the prior years, the amount of debt, the capital assets, and the retained earnings. Auditors will note areas where the utility can improve its accounting on finance methods. It also possible to note areas where bond issues and asset management

Table 12-4 Summary of a proposed budget showing prior expenditures and proposed expenditures for the coming year

Account Number	Account Description	2007 Actuals	2008 Last Year's Actuals	2009 Original Budget	2009 Modified Actual	2009 Changes to Line Items
403-3801-538.31-10	Salaries & Benefits	$388,302	$425,013	$471,512	$427,077	$$44,435
	Professional Services	14,876	5,879	26,271	11,300	14,971
403-3801-538.31-30	Consultant Engineers	20,218	33,109			0
403-3801-538.32-10	Accounting & Auditing	1,132	867	960	0	960
403-3801-538.34-10	Contractual Services	3,661	1,190	1,056	1,040	16
403-3801-538.40-10	Training & Per Diem	1,306	2,090	1,988	2,000	(12)
403-3801-538.40-30	Expense Account	601	603	600	600	0
403-3801-538.41-10	Telephone	387	636	780	648	132
403-3801-538.43-20	Electricity	8,076	7,133	9.966	9,966	0
403-3801-538.44-10	Equipment Rentals	274	255	255	255	0
403-3801-538.44-20	Trailer Rentals	900	900	900	900	0
403-3801-538.44-30	Misc. Rentals	977	992	1,010	1,094	(84)
403-3801-538.45-10	Insurance Coverage	6,271	6,139	5,142	5,553	(411)
403-3801-538.46-10	Equipment Maintenance	15,549	8,113	8,000	10,000	(2,000)
403-3801-538.46-30	Vehicle Maintenance	13,111	15,747	12,000	12,000	0
403-3801-538.46-50	Grounds Maintenance	21,569	18,573	18,000	18,284	(284)
403-3801-538.46-60	Utility Maintenance	14,498	10,040	52,000	60,000	(8,000)
403-3801-538.49-30	Permits & License Fees	4,149	5,543	4,984	20,000	(15,016)
403-3801-538.51-10	Office Supplies	145	150	150	150	0
403-3801-538.52-10	Gasoline	8,782	11,125	15,848	11,662	4,186
403-3801-538.52-20	Misc. Supplies	1,037	–24,034	7,000	7,000	0
403-3801-538.52-50	Uniforms	540	559	750	750	0
403-3801-538.52-60	Building Supplies	275	0	0	0	0
403-3801-538.62-10	Buildings	0	0	162,488	0	162,488
403-3801-538.64-00	Machines & Equipment	46,919	(1,843.00)	0	0	0
403-3801-538.64-20	Vehicles	23,219	0	0	0	0
403-3801-538.64-30	Capital	23,700	1,843	3,500	0	3,500
403-3801-538.71-10	Principal	0	0	16,446	17,183	(737)
403-3801-538.72-10	Interest	4,029	5,694	2,666	1,930	736
403-3801-538.91-10	Transfer to General Fund for purchases and services	0	0	30,000	0	30,000
403-3801-538.99-10	Contingency	0	26,030	181,195	0	181,195
	TOTAL OPERATING	$624,503	$562,346	$1,035,467	$619,392	$416,075

may be improved. Ultimately the audit provides information about whether the financial operations are in compliance with generally accepted accounting principles, they adequately present the financial condition of the utility, and appropriate laws and regulations have been complied with. It also provides some indication of whether expenses appear to be appropriate for labor, energy, chemicals, and other expenses of the utility.

The utility should have some form of vision statement about where it desires to go over a period of time, and the audit will often consider whether the expenditures and programs from the year being audited have complied, or are moving toward compliance, with those goals and objectives. There are alternatives that might be useful to improve compliance or there may be policies that need to be adjusted to meet the goals of the governing board.

Formal Public Relations Programs

Although water distribution personnel play an integral part in creating a favorable image for the utility, there are many other facets of public relations that contribute to the total picture. The size of a water utility may influence the formality or scope of a public relations program, but not its importance.

Large utilities may confront political and environmental issues that affect a cross section of community interest. They may face a high volume of maintenance and repairs, billing, or collection problems; rate-allocation difficulties; or growth patterns that strain current treatment and personnel levels. A larger utility, therefore, will usually maintain a full-time public relations staff whose primary concern is to promote the company's image within the community. This staff will dispense information to the public and work closely with the media.

Smaller utilities are not immune to political, environmental, or community issues, but by necessity they may have to operate with a one- or two-person customer relations staff. In addition, there will likely be a manager who deals with local government, community representatives, and the public in general. Some functions of a larger utility's public relations efforts are discussed in the following sections.

Customer Service

Customer service representatives answer customer questions and handle complaints. They respond to telephone calls and complete the resulting paperwork. Some problems can be solved within minutes, whereas others are more complicated or involve intricate billing problems. Complex or particularly sensitive problems may be referred to a supervisor. In general, customer service representatives are well versed in telephone etiquette, active listening techniques, utility procedures, and persuasive speaking.

Public Information

Public relations specialists dispense information to the community. Their primary goal is to project a favorable image of the organization to the public. Public speaking engagements, participation in civic or professional clubs, and the creation of public service or special school projects consume much of the public information expert's time. Utilities that are conscious of their community image will often support local television, particularly educational or public service channels.

The public information specialist may also create and distribute literature explaining utility operations and policies or brochures giving useful water-related information to consumers.

Media Relations

Public relations personnel generally handle most communications with newspapers, magazines, radio, and television. An ongoing program highlights utility-sponsored projects, upper-management personnel changes, conservation efforts, or any newsworthy information. Bond issues, rate increases, special projects and how they affect homeowners and businesses, or emergency situations are explained to the public through the news media.

A press conference is an interview held for reporters. It is conducted by a utility spokesperson, usually a public relations or media relations expert, the general manager, or the manager's designate. A press conference is usually called to explain an emergency that has some impact on the community as a whole or to make some type of announcement. Media relations experts know what to say and how to say it in order to put the utility in the best possible light.

For more information about these topics, consult resources on effective utility management (www.watereum.org).

Study Questions

1. Who should inspect the work of a water storage painting contractor?
   a. The water utility supervisor
   b. A water operator or operators responsible for the area the tank is located in
   c. At least two competing painting contractors
   d. A qualified third party

2. The most expensive part of a pipe installation is/are the
   a. pipe fittings.
   b. valves.
   c. excavation.
   d. engineers' design.

3. Where do good public relations start for a water utility?
   a. With well-maintained fire hydrants
   b. With meaningful advertising
   c. Via public relations campaigns
   d. Through dedicated service-oriented employees

4. One of the most important functions and the major part of the work day for a manager of a water utility or distribution system is involved in
   a. organizing.
   b. planning.
   c. directing.
   d. controlling.

5. _____ is the most common problem in the workplace.
   a. Miscommunication
   b. Absenteeism
   c. Disobedience
   d. Forgetfulness
6. A _____ is a guideline that employees are expected to follow.
   a. regulation
   b. policy
   c. law
   d. code
7. How is the rate of annual breakages in pipeline calculated?
8. List four typical functions of a supervisor.
9. What is the term for a financial review that comes after the budget year has been completed in which an external accounting group evaluates whether the revenues and expenditures were appropriately categorized, money was maintained in the appropriate accounts, and revenues and expenses were fully accounted for and appropriately spent?

Chapter 13
Additional Study Questions

These questions are provided to help those studying for higher-level certification exams to determine if they should review material included in the *WSO Water Distribution, Grades 1 & 2* book.

Pipe

1. Which of the following is the best or first choice corrosion-control method(s) to protect pipe?
   a. Use of noncorrosive metals and/or mechanical coatings
   b. Chemical protective coatings
   c. Electrical control by using cathodic protection
   d. Use of metallic coatings such as zinc or aluminum

2. Which type of pipe material is prestressed?
   a. Steel cylinder
   b. Ductile iron
   c. Asbestos–cement
   d. C900 PVC

3. When a metal is galvanized, it is coated with
   a. zinc.
   b. aluminum.
   c. aluminum oxide.
   d. aluminum hydroxide.

4. Which of the following types of pipe is most prone to sliding out of a push-on joint if not firmly restrained?
   a. Asbestos–cement pipe
   b. Plastic pipe
   c. Ductile-iron pipe
   d. Cast-iron pipe

5. How is pipe strength expressed?
   a. Hydrostatic potential
   b. Durability is psi
   c. Tensile and flexural strength
   d. Baud units

6. The resistance of a material to longitudinal pulling forces before it breaks is called
   a. flexural strength.
   b. shear strength.
   c. ductile strength.
   d. tensile strength.

7. Which of the following is a type of joint for concrete piping?
   a. Expansion joint
   b. Push-on joint
   c. Bell-and-spigot type
   d. Flanged joint

8. Polyvinyl chloride piping using solvent weld joints is most appropriate
   a. only for small lines.
   b. for high-pressure applications.
   c. where flexibility is required.
   d. where valves or fittings are to be attached.

9. Which type of pipe joint is available in both bolted and boltless flexible pipe joint designs?
   a. Ball-and-socket joint
   b. Push-on joint
   c. Grooved joint
   d. Shouldered joint

Water Main Installation and Rehabilitation

1. How can a distribution operator tell if a new pipe with no obvious cracks or chips is good before it is placed in the trench for installation?
   a. By conducting a pressure test
   b. By gently tapping the length of the pipe with a hammer
   c. By attaching a sonic meter at one end and listening at other end with headphones
   d. By placing a sonic meter at one end and an oscilloscope at the other end

2. Which of the following is the recommended trench width for a 42-in. (1,067-mm) diameter ductile-iron pipe?
   a. 58 in. (1,473 mm)
   b. 60 in. (1,524 mm)
   c. 66 in. (1,676 mm)
   d. 72 in. (1,829 mm)

3. The first layer of backfill if compaction is required for a newly installed pipe should come up to
   a. the bottom of the pipe.
   b. one-third of the way to the bottom of the pipe.
   c. the centerline of the pipe.
   d. the top of the pipe.

4. If special bedding material is required by the design engineer due to poor local soil conditions, the material should not contain granular material greater than
   a. ¼ in. (6 mm).
   b. ⅜ in. (10 mm).
   c. ½ in. (13 mm).
   d. 1 in. (25 mm).

Backfilling, Main Testing, and Installation Safety

1. What is the definition of leakage as it relates to pipeline testing?
2. When pressure testing a new pipeline, what is the usual minimum test pressure?
3. Flushing is usually conducted at what minimum velocity?
4. What are the methods for disinfecting pipelines?
5. How long should the chlorine contact period be before flushing and testing?

Water Services

1. Which type of service pipe material is no longer acceptable for drinking water?
   a. Brass
   b. Lead
   c. Galvanized wrought iron
   d. Copper

2. Why should PVC pipe not be used when soil is contaminated with petroleum products?

3. What is the valve used to connect service lines to water mains?

4. What methods are used to thaw frozen water service lines?

5. Who is responsible for repairing water service lines?

Valves

1. Which type of valve would be particularly useful for throttling the flow of corrosive liquids?
   a. Diaphragm valve
   b. Butterfly valve
   c. Gate valve
   d. Pinch valve

2. When a pressure-reducing and a pressure-sustaining valve are used in combination, one valve can keep a constant ________ pressure even with fluctuating demand, while the other valve holds the pressure at a minimum predetermined ________.
   a. upstream; flow
   b. upstream; pressure
   c. downstream; flow
   d. downstream; pressure

3. What are cautions when using hydraulic actuators to operate pump station valves?

4. How should valve vaults be drained?

Fire Hydrants

1. What are some of the important concerns in fire hydrant installation?

2. What is the maximum pressure for testing for fire hydrant leaks?

3. When should a hydrant be repaired or replaced?

4. How often should fire hydrants be inspected?

5. What operation should inspection procedures include?

Water Storage

1. Which of the following water storage facilities is most likely to contain nonpotable water?
   a. Standpipe
   b. Buried storage tanks
   c. Emergency storage tanks
   d. Elevated tanks using a riser; one way in and one way out

2. What is the proper detention time for disinfecting a water storage tank with water that is mixed with hypochlorite already in the tank such that the free chlorine is 10 mg/L after proper detention time is complete?
   a. 6 hours
   b. 8 hours
   c. 12 hours
   d. 24 hours

3. What is the available chlorine concentration used for disinfecting a water storage tank using a method that involves spraying or painting of all the interior tank surfaces?
   a. 50 mg/L
   b. 100 mg/L
   c. 200 mg/L
   d. 250 mg/L

4. Why are altitude valves used on water storage tanks?
   a. To allow water to pass into and out of the tank as pressure fluctuates
   b. To stop the flow of water into the tank when it is full
   c. To allow overflow water to flow out of the tank
   d. To shut the flow of water to the tank off for maintenance and inspection

Pumps and Pumping Stations

1. The height a liquid can be raised vertically by a given pressure is called
   a. pressure head.
   b. total head.
   c. velocity head.
   d. pump head.

2. A split-case pump has three impellers. Which type of multistage pump is this?
   a. One stage
   b. Two stage
   c. Three stage
   d. Six stage

3. A split-case pump has two equal smaller impellers placed on either side of two equally sized large impellers. How many stages does this pump have?
   a. One stage
   b. Two stages
   c. Four stages
   d. Eight stages

4. A pump loses its prime because the suction line has an air pocket. Which of the following is the best solution?
   a. Check the pump's amperage and be sure the pump's strainer is clean.
   b. Clean or repair the priming unit.
   c. Open suction piping air bleed-off valves.
   d. Check the external water seal unit.

5. The lowest pressure point in the pump is the
   a. center of the impeller.
   b. outermost part of the impeller.
   c. suction side of the pump.
   d. discharge side of the pump.

6. The shaft's main function is to transmit ________ from the motor to the impeller.
   a. centrifugal force
   b. torque
   c. kinetic energy
   d. thrust

7. Which of the following needs to be determined first when designing a pump station?
   a. Discharge requirements
   b. Head requirements
   c. Power requirements
   d. Capacity requirements

Basic Chlorination

1. Which chemical oxidant would be most effective for controlling biological growth?
   a. Chloramines
   b. Chlorine
   c. Ozone
   d. Potassium permanganate

2. What is the most common hypochlorinator problem?
   a. Clogged equipment
   b. Cracked pump head
   c. Corrosion
   d. Broken plunger

3. Which reaction will take longer with chlorine?
   a. Hydrogen sulfide
   b. Ammonia
   c. Organic material
   d. Calcium sulfide

4. Which organisms have the greatest resistance to chlorine?
   a. Viruses
   b. Bacteria
   c. Nematodes
   d. *Giardia* cysts

5. If secondary chlorine is added to a distribution system that uses chloramines, the chlorine may be added to a desired chlorine to ammonia-nitrogen ratio of
   a. 3.0:1 to 3.5:1.
   b. 3.5:1 to 4.0:1.
   c. 4.0:1 to 4.5:1.
   d. 4.5:1 to 5.0:1.

6. What is the most probable solution if sulfur bacteria are causing corrosion in the distribution system?
   a. Minimization of pump activity
   b. Acidification and cleaning
   c. Optimization of coagulation, flocculation, and filtration
   d. Routine use of disinfectant and penetrant

Backflow Prevention and Cross-Connection Control

1. What is the height limit to which siphoned water can be lifted at sea level?
   a. 22.4 ft (6.8 m)
   b. 32.0 ft (9.8 m)
   c. 33.9 ft (10.3 m)
   d. 34.0 ft (10.4 m)

2. What is the best protection from backflow when filling a tank truck from a fire hydrant?

3. How often and by whom are backflow prevention devices tested in most cross-connection control programs?

4. Is a fire pumper truck connected to a hydrant for firefighting purposes considered a cross-connection?

5. What type of backflow prevention device is mainly used for automatic landscape sprinkler systems?

Study Question Answers

Chapter 1 Answers

1. **d.** No more than 5%
2. **d.** 4 log
3. **a.** population served is >3,300 people.
4. **d.** 100,000 people.
5. **b.** <2.0 mg/L
6. **d.** Chloramine disinfection
7. **b.** is a Tier 2 violation requiring public notification within 30 days.
8. **b.** meeting disinfection limits.
9. **d.** the treated water quarterly running average TOC is less than 2.0 mg/L.
10. Safe Drinking Water Act (SDWA)
11. 1.3 mg/L
12. 15 pCi/L
13. Maximum residual disinfectant level (MRDL)
14. 10 µg/L

Chapter 2 Answers

1. **b.** Capacity (flow rate)
2. **d.** 310 mhp

$$\text{Equation: Motor hp} = \frac{\text{whp}}{(\text{Motor effic.})(\text{Pump effic.})}$$

$$= \frac{200\ \text{whp}}{(88\%/100\%\ \text{Motor effic.})(74\%/100\%\ \text{Pump effic.})}$$

$$= \frac{200\ \text{whp}}{(0.88\ \text{Motor effic.})(0.74\ \text{Pump effic.})}$$

= 307 mhp, round to **310 mhp**

3. **c.** 1,250 gpm

First, convert the diameter of the pipe from inches to feet:

$$\text{Diameter, ft} = (10.0\text{ in.})(1\text{ ft}/12\text{ in.}) = 0.833\text{ ft}$$

Next, calculate the pipe's cross-sectional area in square feet:

$$\text{Area, ft}^2 = (0.785)(\text{Diameter, ft})^2$$

$$\text{Area, ft}^2 = (0.785)(0.833\text{ ft})(0.833\text{ ft}) = 0.5447\text{ ft}^2$$

Next, find the flow in the pipe in cfs:

$$\text{Flow, cfs} = (\text{Area, ft}^2)(\text{Velocity, ft/sec})$$

$$\text{Flow, cfs} = (0.5447\text{ ft}^2)(5.10\text{ ft/sec}) = 2.778\text{ cfs}$$

Last, determine the reading on the flowmeter in gpm:

$$\text{Flow, gpm} = (2.778\text{ cfs})(7.48\text{ gal/ft}^3)(60\text{ sec/min}) = 1{,}247\text{ gpm, round to }\mathbf{1{,}250\text{ gpm}}$$

4. **b.** 350 gal

First, determine the capacity of the tank in gallons:

$$\text{Volume, gal} = (0.785)(\text{Diameter, ft})^2(\text{Height, ft})(7.48\text{ gal/ft}^3)$$

$$\text{Volume, gal} = (0.785)(84.0\text{ ft})(84.0\text{ ft})(24.25\text{ ft})(7.48\text{ gal/ft}^3) = 1{,}004{,}712\text{ gal}$$

Next, convert number of gallons to million gallons:

$$1{,}004{,}712\text{ gal} \div 1{,}000{,}000 = 1.0047\text{ mil gal}$$

Using the "pounds" formula, determine the chlorine pounds needed:

$$\text{Chlorine, lb} = (\text{Volume, mil gal})(\text{Dosage, mg/L})(8.34\text{ lb/gal})$$

$$\text{Chlorine, lb} = (1.0047\text{ mil gal})(50.0\text{ mg/L})(8.34\text{ lb/gal}) = 418.96\text{ lb}$$

$$\text{NaOCl solution, gal} = [(\text{Chlorine, lb})(100\%)] \div [(\text{NaOCl, lb/gal})(\text{Hypochlorite, }\%)]$$

$$= [(418.96\text{ lb})(100\%)] \div [(9.59\text{ lb/gal})(12.5\%)]$$

$$= 349.5\text{ gal, round to }\mathbf{350\text{ gal}}$$

5. **a.** 43 oz of NaOCl

First, find the length (in feet) of water in the casing:

Length of water-filled casing = Depth of well – Depth of water to top of casing

Length of water-filled casing = 210 ft – 91 ft = 119 ft

Then convert the diameter from inches to feet:

$$\text{Diameter, ft} = 14.0\text{ in.} \div 12\text{ in./ft} = 1.1667\text{ ft}$$

Next, determine the volume in gallons of water in the well casing using the following formula:

$$\text{Volume, gal} = (0.785)(\text{Diameter, ft})^2(\text{Length, ft})(7.48\text{ gal/ft}^3)$$

$$\text{Volume, gal} = (0.785)(1.1667\text{ ft})(1.1667\text{ ft})(119\text{ ft})(7.48\text{ gal/ft}^3) = 951.12\text{ gal}$$

Next, determine the number of mil gal:

$$\text{mil gal} = 951.12 \text{ gal} \div 1{,}000{,}000 = 0.000951 \text{ mil gal}$$

Lastly, using the "pounds" formula, calculate the number of lb of sodium hypochlorite:

$$\text{Sodium hypochlorite, lb} = \frac{(0.000951 \text{ mil gal})(50.0 \text{ mg/L})(8.34 \text{ lb/gal})}{(12.5\% \text{ available chlorine}/100\%)}$$

$$= 3.173 \text{ lb}$$

Next, find the number of ounces sodium hypochlorite (NaOCl):

$$\text{Number of ounces} = [(3.173 \text{ lb})(128 \text{ oz/gal})] \div [9.50 \text{ lb/gal}]$$
$$= 42.75 \text{ oz, round to } \mathbf{43\ oz}$$

6. **b.** 17 ft, therefore NPSHA < NPSHR so cavitation should occur

First determine the atmospheric pressure in feet. Know that 1.09 ft/in. Hg. Thus:

$$\text{Atmospheric pressure} = (29.8 \text{ in. Hg})(1.11 \text{ ft/in. Hg}) = 33.078 \text{ ft}$$

Next determine the NPSHA:

$$\text{NPSHA} = \text{AP, ft} - \text{SSL, ft} - \text{Hf, ft} - \text{VP, ft}$$
$$\text{NPSHA} = 33.078 \text{ ft} - 15.1 \text{ ft} - 0.61 \text{ ft} - 0.50 \text{ ft}$$
$$\text{NPSHA} = 16.868 \text{ ft, round to } 17 \text{ ft}$$

Therefore: **NPSHA 17 ft < NPSHR 18.4 ft, so cavitation should occur.**

7. **d.** 14.8 mA

$$\text{Current process reading} = \frac{(\text{Live signal, mA} - 4 \text{ mA offset})(\text{Maximum capacity})}{16 \text{ milliamp span}}$$

Substitute known values and solve:

$$22.89 \text{ ft (Storage tank level)} = \frac{(\text{Live signal, mA} - 4 \text{ mA offset})(34.0 \text{ ft Maximum level})}{16 \text{ mA}}$$

Rearrange the equation to solve for live signal in mA:

$$\text{Live signal mA} - 4 \text{ mA offset} = \frac{(22.89 \text{ ft})(16 \text{ mA})}{34.0 \text{ ft}}$$

$$\text{Live signal mA} = \frac{(22.89 \text{ ft})(16 \text{ mA})}{34.0 \text{ ft}} + 4 \text{ mA offset}$$

$$= 10.77 \text{ mA} + 4 \text{ mA offset}$$

$$= 14.77 \text{ mA, round to } \mathbf{14.8\ mA}$$

8. **d.** discharge side
9. Head loss
10. chlorine dosage (mg/L) = chlorine demand (mg/L) + chlorine residual (mg/L)
11. By multiplying the motor and pump efficiencies together
12. feed rate = (dosage)(flow rate)(conversion factor)

Chapter 3 Answers

1. **b.** 20 psi (138 kPa).
2. **d.** Tree system
3. **b.** 2.5–4.0 times
4. **d.** to complete a grid.
5. **b.** 2–4 ft/sec (0.6–1.2 m/sec)
6. **b.** fire demand.
7. **b.** tree system
8. Groundwater system
9. Rural water system
10. Grid system
11. Maximum day demand

Chapter 4 Answers

1. **c.** proportionately.
2. **a.** Grid system
3. **c.** potential energy.
4. **a.** 1; 2
5. **d.** fluids in motion and at rest.
6. **a.** Power failure shutting down a pump suddenly
7. **d.** Velocity
8. **b.** water hammer
9. Static pressure
10. 8 mm
11. $Q = A \times V$, or quantity of water = cross-sectional area of pipe × velocity of flow

Chapter 5 Answers

1. **c.** Transmitter, transmission channel, and receiver
2. **a.** digital signals.
3. **b.** Signal conditioners, actuators, and control elements
4. **a.** Thermistor
5. **d.** RTUs, communications, master station, and HMI
6. **a.** 4–20 mA DC
7. Automatic

8. Direct manual
9. Voltage
10. Solenoid
11. Polling

Chapter 6 Answers

1. **d.** 5–10
2. **a.** Wound-rotor induction motor and a controller
3. **b.** Voltage relay
4. **a.** Thermal-overload relay
5. **c.** Frequency relay
6. **d.** Differential relay
7. **d.** Decreasing the size of power lines and transformers
8. 5%
9. Increasing system efficiency, spreading the pumping load more evenly throughout the day, and reducing power-factor charges
10. 1.0
11. Near at hand and ready for use
12. Phase imbalance

Chapter 7 Answers

1. **a.** 1–2 in. (21–51 mm)
2. **d.** Provide proper alignment and support for the meter
3. **a.** 5–10%
4. **a.** in a horizontal plane
5. **c.** gallons or cubic feet.
6. **b.** Days of the week on which the meter may be read
7. Register
8. The meter should be located immediately after the point where the service pipe enters through the floor or wall. If meters are to be read directly, the location must be kept relatively clear to allow convenient reading. If meters are to be furnished with remote reading devices, the location should still allow for reasonable access.

Chapter 8 Answers

1. **d.** direct currents.
2. **c.** galvanic anodes.
3. **a.** Specifying extra thickness for pipe walls
4. **d.** air purging
5. **b.** unauthorized persons cannot remove them by hand.
6. Listening surveys, correlator method, statistical noise analyzer
7. Use a well-planned directional flushing program to restore chlorine residuals then retest to determine if additional steps (shock disinfection) may be necessary.

8. Backpressure
9. Galvanic cell
10. Any four of the following: system expansion that does not provide additional feeder main; new water lines installed at elevations higher or lower than the original system; additional customer services added to existing mains; unintentionally closed or partially closed valves; undetected leaks in mains or services; changes to water storage tanks; reductions in pipe capacity due to corrosion, pitting, tuberculation, sediment deposits, or slime growth

Chapter 9 Answers

1. **a.** rad.
2. **a.** negative.
3. **b.** Gas chromatograph
4. **d.** Obtain the customer's name and address at the beginning of the discussion.
5. 24 hours is generally needed for incubation, but some tests may be completed in 18 hours.
6. If ammonia is present, chloramines form, increasing the residual; as more is added, these chloramines are partially destroyed until the breakpoint is reached and the residual rises again.
7. The value changes, so field tests are needed for accurate results.
8. Completed test
9. Approximately 5 minutes

Chapter 10 Answers

1. **c.** modeling and analysis.
2. **d.** Base data
3. **c.** work order
4. **a.** Geographic information system
5. **b.** land records data.
6. **c.** Maintenance management system
7. SCADA, maintenance, outage management, work management, emergency response, engineering and design, laboratory management
8. Physical location
9. Corrective maintenance and preventive maintenance
10. A laboratory information management system

Chapter 11 Answers

1. **b.** Ricin
2. **a.** 10 ppm.
3. **b.** Paralyzes the respiratory system
4. **b.** 15 minutes.
5. **a.** emergency response plan

6. Cyber, physical, chemical, biological
7. Vulnerability assessment
8. The US Department of Energy
9. It should be suitable for physical and environmental conditions, it should have no gaps in coverage, and it should present layers of protection.

Chapter 12 Answers

1. **d.** A qualified third party
2. **c.** excavation.
3. **d.** Through dedicated service-oriented employees
4. **b.** planning.
5. **a.** Miscommunication
6. **b.** policy
7. Number of breaks ÷ 100 miles of pipeline mains
8. Any four of the following: planning, budgeting, organizing, staffing, directing, encouraging communication, controlling, managing time, and decision making
9. Annual audit

Chapter 13 Answers

Pipe

1. **a.** Use of noncorrosive metals and/or mechanical coatings
2. **a.** Steel cylinder
3. **a.** zinc.
4. **b.** Plastic pipe
5. **c.** Tensile and flexural strength
6. **d.** tensile strength.
7. **c.** Bell-and-spigot type
8. **a.** only for small lines.
9. **a.** Ball-and-socket joint

Water Main Installation and Rehabilitation

1. **b.** By gently tapping the length of the pipe with a hammer
2. **c.** 66 in. (1,676 mm)
3. **c.** the centerline of the pipe.
4. **d.** 1 in. (25 mm).

Backfilling, Main Testing, and Installation Safety

1. Leakage is the amount of water added to a full pipeline to maintain pressure within a 5-psi (34-kPa) range.
2. 150 psi (1,000 kPa) or 1.5 times the operating pressure for 30 minutes

3. 2.5 ft/sec for two or three complete changes of water within the pipeline
4. Continuous feed, slug, and tablet methods
5. Usually 24 hours but longer if unsanitary conditions are encountered

Water Services

1. **b.** Lead
2. Because these substances can soften and weaken the pipe and penetrate into the water supply
3. Corporation stop, usually with Mueller threads
4. Electrical using a welding unit and hot water
5. Usually the property owner, but some systems have that responsibility

Valves

1. **d.** Pinch valve
2. **d.** downstream; pressure
3. Deposits may build up in water-operated valve cylinders and prevent proper operation, and if system pressure is used, operation may be inhibited under low-pressure conditions.
4. Do not drain to storm drains or sanitary sewers because of possible contamination; instead it is usually advisable to drain to underground adsorption pits.

Fire Hydrants

1. Location, footing and blocking, drainage, color
2. 150 psi (1,000 kPa)
3. If it cannot be operated by one person using a standard 15-in. wrench
4. At least annually
5. Opening the hydrant to maximum flow

Water Storage

1. **c.** Emergency storage tanks
2. **d.** 24 hours
3. **c.** 200 mg/L
4. **b.** To stop the flow of water into the tank when it is full

Pumps and Pumping Stations

1. **a.** pressure head.
2. **b.** Two stage
3. **b.** Two stages
4. **c.** Open suction piping air bleed-off valves.
5. **a.** center of the impeller.
6. **b.** torque
7. **b.** Head requirements

Basic Chlorination

1. **b.** Chlorine
2. **a.** Clogged equipment

3. **c.** Organic material
4. **c.** Nematodes
5. **d.** 4.5:1 to 5.0:1.
6. **d.** Routine use of disinfectant and penetrant

Backflow Prevention and Cross-Connection Control

1. **c.** 33.9 ft (10.3 m)
2. An air gap on the fill line at least two times the inside diameter of the fill line
3. Annually by a certified tester
4. Most systems don't consider this a cross-connection, but it is possible for the end of the fire hose to lie in a puddle of nonpotable water and potentially provide a route for contamination if backflow exists.
5. A reduced pressure zone (RPZ) backflow preventer

References

Abbaszadegan, M. and A. Alum. 2004. Microbiological Contaminants and Threats of Concern. In Water Supply Systems Security, ed. L. W. Mays. New York: McGraw-Hill.

Booth, R., A. Bowman, F. Gist, and J. Ringold. 2004. Security Hardware and Surveillance Systems for Water Supply Systems. In *Water Supply Systems Security*, ed. L. W. Mays. New York: McGraw-Hill.

Clark, R. M. and R. A. Deininger. 2001. *Minimizing the Vulnerability of Water Supplies to Natural and Terrorist Threats.* In Proceedings of the American Water Works Association IMTech Conference, Atlanta, GA.

Collins J. 2001. *Good to Great: Why Some Companies Make the Leap...and Others Don't.* New York: HarperBusiness.

Diamond, L.E. 2007. *Team Building That Gets Results.* Naperville, IL.: Some Books Inc.

Disinfecting Water Mains. ANSI/AWWA Standard 651. Denver, CO: American Water Works Association (AWWA).

Distribution System Requirements for Fire Protection. AWWA Manual M31. Denver, CO: AWWA.

Emergency Planning for Water Utilities. AWWA Manual M19. Denver, CO: AWWA.

FWPCOA. 1997. *Supervision.* St. Lucie, FL: Florida Water and Pollution Control Operator's Association.

Grayman, W. M., R. M. Clark, B. L. Harding, M. Maslia, and J. Aramini. 2004a. Reconstructing Historical Contamination Events. In *Water Supply Systems Security*, ed. L. W. Mays. New York: McGraw-Hill.

Grayman, W. M., R. A. Deininger, R. M. Males, and R. W. Gullick. 2004b. Source Water Early Warning Systems. In *Water Supply Systems Security*, ed. L. W. Mays. New York: McGraw-Hill.

International Life Sciences Institute, Risk Science Institute. 1999. *Early Warning Monitoring to Detect Hazardous Events in Water Supplies.* Washington, DC: ILSI Press.

Johnson, A., and R. Gabriel. 2009. Preparedness Reduces Risk. *Opflow*, 35(9): 14–17.

Manitoba Office of Drinking Water. 2009. *Emergency Planning for Water Utilities in Manitoba.* Available at https://www.gov.mb.ca/waterstewardship/odw/reg-info/operations-monitor/emergency_water_utilities_mar09.pdf.

Mays, L. W., ed. 2004. *Water Supply Systems Security*. New York: McGraw-Hill.

National Fire Protection Association. 2007. *NFPA 1142: Standard on Water Supplies for Suburban and Rural Fire Fighting*. Quincy, MA: NFPA.

Nitrification Prevention and Control in Drinking Water. AWWA Manual M56. Denver, CO: AWWA.

Oberoi, K. *Operational Guide to AWWA Standard G200: Distribution Systems Operation and Management*. Denver, CO: American Water Works Association.

Panguluri, S., W. R. Phillips Jr., R. M Clark. 2004. Cyber Threats and IT/SCADA System Vulnerability. In *Water Supply Systems Security*, ed. L. W. Mays. New York: McGraw-Hill.

Phillips, W. R. Jr. Typical Water/Wastewater Utility's Business and SCADA Infrastructure and Network Connectivity. Copyright 2009 by CH2MHill. Reprinted with Permission.

Ralston, S., and J. Ginley. 2010. In Deep Water. *American City and County*, p. 30–43.

Rice, E. W., Baird, R. B., and L. S. Eaton, eds. 2012. *Standard Methods for the Examination of Water and Wastewater*. American Public Health Association, AWWA, and Water Environment Federation.

Risk and Resilience Management of Water and Wastewater Systems. ANSI/AWWA Standard J100. Denver, CO: AWWA.

Smith, C., ed. 2005. Water *Distribution System Assessment Workbook*. Denver, CO: American Water Works Association.

Taylor, F. 1911. *Principles of Scientific Management*. New York: Harper & Brothers.

US Department of Energy, President's CIP Board. 2002. 21 *Steps to Improve Cyber Security of SCADA Networks*. Available at http://www.oe.netl.doe.gov/docs/prepare/21stepsbooklet.pdf.

US Environmental Protection Agency. 2002. *Vulnerability Assessment Fact Sheet*. Office of Water (4601M), EPA 816-F-02-025. Washington, DC: USEPA.

Water Audits and Loss Control Programs. AWWA Manual M36. Denver, CO: AWWA.

Water Meters: Selection, Installation, Testing and Maintenance. AWWA Manual M6. Denver, CO: AWWA.

Water Treatment Plant Operation and Management. ANSI/AWWA Standard G100. Denver, CO: AWWA.

Glossary

absence See *negative sample.*

air purging A procedure to clean mains less than 4 in. (100 mm) in diameter, in which air from a compressor is mixed with the water and flushed through the main.

analog Continuously variable, as applied to signals, instruments, or controls. Compare with *digital.*

annual audit A financial review that comes after the budget year has been completed in which an external accounting group evaluates whether the revenues and expenditures were appropriately categorized, money was maintained in the appropriate accounts, and revenues and expenses were fully accounted for and appropriately spent. Also known as *comprehensive annual financial report* or *CAFR*.

arterial-loop system A distribution system layout involving a complete loop of arterial mains (sometimes called trunk mains or feeders) around the area being served, with branch mains projecting inward. Such a system minimizes dead ends.

automated mapping/facility management/geographic information system (AM/FM/GIS) A computerized system for collecting, storing, and analyzing water system components for which geographic location is an important characteristic.

automatic control A system in which equipment is controlled entirely by machines or computers, without human intervention, under normal conditions.

automatic meter reading (AMR) Any of several methods of obtaining readings from customer meters by a remote method. Methods that have been used include transmitting the reading through the telephone system, through the electric power network, through water lines (via sound transmission), through cable TV wiring, and by radio.

average day demand The total system water use for 1 year divided by 365 days in a year.

backfill (1) The operation of refilling an excavation, such as a trench, after the pipeline or other structure has been placed into the excavation. (2) The material used to fill the excavation in the process of backfilling.

backflow A hydraulic condition, caused by a difference in pressures, in which non-potable water or other fluids flow into a potable water system.

backpressure A condition in which a pump, boiler, or other equipment produces a pressure greater than the water supply pressure.

backsiphonage A condition in which the pressure in the distribution system is less than atmospheric pressure, which allows contamination to enter a water system through a cross-connection.

brake horsepower The power supplied to a pump by a motor. Compare with *water horsepower* and *motor horsepower*.

breakpoint The point at which the chlorine dosage has satisfied the chlorine demand.

breakpoint chlorination The addition of chlorine to water until the chlorine demand has been satisfied and free chlorine residual is available for disinfection.

brushes Graphite connectors that rub against the spinning commutator in an electric motor or generator, connecting the rotor windings to the external circuit.

***C* × *T* value** The product of the residual disinfectant concentration *C*, in milligrams per liter, and the corresponding disinfectant contact time *T*, in minutes. Minimum *C* × *T* values are specified by the Surface Water Treatment Rule as a means of ensuring adequate kill or inactivation of pathogenic microorganisms in water.

calcium carbonate Scale-forming substance in water.

cathodic protection An electrical system for preventing corrosion to metals, particularly metallic pipe and tanks.

chlorine demand The quantity of chlorine consumed by reaction with substances in water.

coliform bacteria A group of bacteria predominantly inhabiting the intestines of humans or animals but also occasionally found elsewhere. Presence of the bacteria in water is used as an indication of fecal contamination (contamination by human or animal wastes).

combined chlorine residual The chlorine residual produced by the reaction of chlorine with substances in the water. Because the chlorine is "combined" it is not as effective a disinfectant as free chlorine residual. In water treatment, this usually refers to compounds formed by the combination of chlorine and ammonia.

completed test The third major step of the multiple-tube fermentation method. This test confirms that positive results from the presumptive test are due to coliform bacteria. See also *confirmed test*; *presumptive test*.

confirmed test The second major step of the multiple-tube fermentation method. This test confirms that positive results from the presumptive test are due to coliform bacteria. See also *completed test*; *presumptive test*.

control terminal unit (**CTU**) The receiver in a digital signal system.

corrosion The gradual deterioration or destruction of a substance or material by chemical action, frequently induced by electrochemical processes. The action proceeds inward from the surface.

cross-connection Any arrangement of pipes, fittings, fixtures, or devices that connects a nonpotable system to a potable water system.

current (1) The flow rate of electricity, measured in amperes. (2) In telemetry, a signal whose amperage varies as the parameter being measured varies.

design point The mark on the *H–Q* (head–capacity) curve of a pump characteristics curve that indicates the head and capacity at which the pump is intended to operate for best efficiency in a particular installation.

detention time The average length of time a drop of water or a suspended particle remains in a tank or chamber. Mathematically, it is the volume of water in the tank divided by the flow rate through the tank.

digital Varying in precise steps, as applied to signals or instrumentation and control devices. Compare with *analog*.

direct manual control A type of system control in which personnel manually operate the switches and levers to control equipment from the physical location of the equipment.

direct-wire control A system for controlling equipment at a site by running wires from the equipment to the onsite control panel.

discharge side The outlet, or high-pressure, side of a pump.

disinfection by-products (DBPs) New chemical compounds that are formed by the reaction of disinfectants with organic compounds in water. At high concentrations, many DBPs are considered a danger to human health.

duplexing A type of telemetry in which a single line allows the operator to send the instrument signals that are received at the central location back to a remote site.

dynamic discharge head The difference in height measured from the pump center line at the discharge of the pump to the point on the hydraulic grade line directly above it.

dynamic pressure Pressure that exists in water as moving energy.

dynamic suction head The distance from the pump center line at the suction of the pump to the point of the hydraulic grade line directly above it. Dynamic suction head exists only when the pump is below the piezometric surface of the water at the pump suction. When the pump is above the piezometric surface, the equivalent measurement is dynamic suction lift.

dynamic suction lift The distance from the pump center line at the suction of the pump to the point on the hydraulic grade line directly below it. Dynamic suction lift exists only when the pump is above the piezometric surface of the water at the pump suction. When the pump is below the piezometric surface, the equivalent measurement is called dynamic suction head.

dynamic water system The description of a water system when water is moving through the system.

equilibrium A balanced condition in which the rate of formation and the rate of consumption of a constituent or constituents are equal.

fire flow The rate of flow, usually measured in gallons per minute (gpm) or liters per minute (L/min), that can be delivered from a water distribution system at a specified residual pressure for firefighting. When delivery is to fire department pumpers, the specified residual pressure is generally 20 psi (140 kPa).

free water surface The surface of water that is in contact with the atmosphere.

friction head loss The head lost by water flowing in a stream or conduit as the result of (1) the disturbance set up by the contact between the moving water and its containing conduit and (2) intermolecular friction.

galvanic cell A corrosion condition created when two different metals are connected and immersed in an electrolyte, such as water.

galvanic corrosion A form of localized corrosion caused by the connection of two different metals in an electrolyte, such as water.

gas chromatograph (GC) A technique used to measure the concentration of organic compounds in water.

grid system A distribution system layout in which all ends of the mains are connected to eliminate dead ends.

groundwater system A water system using wells, springs, or infiltration galleries as its source of supply.

head loss The amount of energy used by water in moving from one location to another.

heterotrophic plate count (HPC) A laboratory procedure for estimating the total bacterial count in a water sample. Also known as *standard plate count*, *total plate count*, or *total bacterial count*.

hydraulic grade line (HGL) A line (hydraulic profile) indicating the piezometric level of water at all points along a conduit, open channel, or stream. In an open channel, the HGL is the free water surface.

hydraulics The study of fluids in motion or under pressure.

hydrostatic pressure The pressure exerted by water at rest (for example, in a nonflowing pipeline).

indicator The part of an instrument that displays information about a system being monitored. Generally either an analog or digital display.

indicator organism A microorganism whose presence indicates the presence of fecal contamination in water.

instantaneous flow rate A flow rate of water measured at one particular instant, such as by a metering device, involving the cross-sectional area of the channel or pipe and the velocity of the water at that instant.

line item An expected cost within an expense category.

maintenance management system (MMS) An organized, typically computerized, way for a utility to keep track of its maintenance needs.

maximum contaminant level (MCL) The maximum permissible level of a contaminant in water as specified in the regulations of the Safe Drinking Water Act.

maximum contaminant level goal (MCLG) Nonenforceable health-based goals published along with the promulgation of an MCL. Originally called *recommended maximum contaminant levels (RMCLs).*

maximum day demand The water use during the 24 hours of highest demand during the year.

membrane filter method A laboratory method used for coliform testing. The procedure uses an ultrathin filter with a uniform pore size smaller than bacteria—less than a micron. After water is forced through the filter, the filter is incubated in a special media that promotes the growth of coliform bacteria. Bacterial colonies with a green-gold sheen indicate the presence of coliform bacteria.

meter box A pit-like enclosure that protects water meters installed outside of buildings and allows access for reading the meter. Also known as *meter pit.*

meter pit See *meter box.*

minor head loss The energy losses that result from the resistance to flow as water passes through valves, fittings, inlets, and outlets of a piping system.

MMO–MUG technique An approved bacteriological procedure for detecting the presence or absence of total coliforms.

motor horsepower The horsepower equivalent to the watts of electric power supplied to a motor. Compare with *brake horsepower* and *water horsepower.*

multiple-tube fermentation (MTF) method A laboratory method used for coliform testing that uses a nutrient broth placed in culture tubes. Gas production indicates the presence of coliform bacteria.

multiplexing The use of a single wire or channel to carry the information for several instruments or controls.

negative sample When referring to the multiple-tube fermentation or membrane filter test, any sample that does not contain coliform bacteria. Also known as *absence.*

on–off differential control A mode of controlling equipment in which the equipment is turned fully on when a measured parameter reaches a preset value, then turned fully off when it returns to another preset value.

operations management plan A plan detailing the distribution system's performance objectives, operational practices to achieve the objectives, and performance measures and numeric goals of the plan.

peak-hour demand The greatest volume of water in an hour that must be supplied by a water system during any particular time period, such as a year, to meet customer demand.

per-capita water use The average day demand divided by the number of residents connected to the water system.

performance measure A standard that a system establishes for itself to ensure that it is meeting key objectives.

piezometer An instrument for measuring pressure head in a conduit, tank, or soil, by determining the location of the free water surface.

piezometric surface An imaginary surface that coincides with the level of the water in an aquifer, or the level to which water in a system would rise in a piezometer.

pig Bullet-shaped polyurethane foam plug, often with a tough, abrasive external coating, used to clean pipelines. Forced through the pipeline by water pressure.

pneumatic Operated by air pressure.

policy A guideline that employees are expected to follow.

polling A technique of monitoring several instruments over a single communications channel with a receiver that periodically asks each instrument to send current status.

positive sample In reference to the multiple-tube fermentation or membrane filter test, any sample that contained coliform bacteria. Also known as *presence*.

power (in hydraulics or electricity) The measure of the amount of work done in a given period of time. The rate of doing work. Measured in watts or horsepower.

presence See *positive sample*.

presence–absence (P–A) test An approved bacteriological procedure for the detection of total coliforms. The results are qualitative rather than quantitative.

pressure The force on a unit area of water.

presumptive test The first major step in the multiple-tube fermentation test. The step presumes (indicates) the presence of coliform bacteria on the basis of gas production in nutrient broth after incubation.

proportional control A mode of automatic control in which a valve or motor is activated slightly to respond to small variations in the system, but activated at a greater rate to respond to larger variations.

pulse-duration modulation (PDM) An analog type of telemetry-signaling protocol in which the time that a signal pulse remains on varies with the value of the parameter being measured.

pump center line An imaginary line through the center of a pump.

pump characteristic curve A curve or curves showing the interrelation of speed, dynamic head, capacity, brake horsepower, and efficiency of a pump.

purchased water system A water system that purchases water from another water system and so generally provides only distribution and minimal treatment.

radiation absorption dose (rad) A measure of the dose absorbed by the body from radiation (100 ergs of energy in 1 g of tissue).

radioactivity Behavior of a material that has an unstable atomic nucleus, which spontaneously decays or disintegrates, producing radiation.

receiver (1) The part of a meter that converts the signal from the sensor into a form thatcan be read by the operator; also called the *receiver–indicator*. (2) In a telemetry system, the device that converts the signal from the transmission channel into a form that the indicator can respond to.

receiver–indicator An instrument component that combines the features of a receiver and an indicator.

register That part of the meter that displays the volume of water that has flowed through the meter. Meter registers are generally either of the straight or circular type.

relay An electrical device in which an input signal, usually of low power, is used to operate a switch that controls another circuit, often of higher power.

remote manual control A type of system control in which personnel in a central location manually operate the switches and levers to control equipment at a distant site.

remote terminal unit (RTU) A computer terminal used to monitor the status of control elements, monitor and transmit inputs from instruments, and respond to data requests and commands from the master station.

roentgen equivalent man (rem) A quantification of radiation in terms of its dose effect on the human body; the number of rads times a quality factor.

roentgen equivalent physical (rep) The quantity of radiation (other than X-rays or other generated radiation) that produces in one gram of human tissue ionization equivalent to the quantity produced in air by one roentgen of radiation or X-rays (equivalent to 83.8 ergs of energy).

running annual average (RAA) The average of four quarterly samples at each monitoring location to ensure compliance with the Stage 2 DBPR.

rural water system A water system that has been established to serve widely spaced homes and communities in areas having no available groundwater or having water of very poor quality.

scanning A technique of checking the value of each of several instruments, one after another. Used to monitor more than one instrument over a single channel.

secondary instrumentation Instruments that respond to and display information from primary instrumentation.

semiautomatic control A form of system control equipment in which many actions are taken automatically but some situations require human intervention.

slug method A method of disinfecting new or repaired water mains in which a high dosage of chlorine is added to a portion of the water used to fill the pipe. This slug of water is allowed to pass through the entire length of pipe being disinfected.

solenoid An electrical device that consists of a coil of wire wrapped around a movable iron core. When a current is passed through the coil, the core moves, activating mechanical levers or switches.

static discharge head The difference in height between the pump center line and the level of the discharge free water surface.

static pressure Pressure that exists in water although the water does not flow.

static suction head The difference in elevation between the pump center line and the free water surface of the reservoir feeding the pump. In the measurement of static suction head, the piezometric surface of the water at the suction side of the pump is higher than the pump; otherwise, static suction lift is measured.

static suction lift The difference in elevation between the pump center line of a pump and the free water surface of the liquid being pumped. In a static suction lift measurement, the piezometric surface of the water at the suction side of the pump is lower than the pump; otherwise, static suction head is measured.

static water system The description of a water system when water is not moving through the system.

suction side The inlet, or low-pressure, side of a pump.

supervisory control and data acquisition (SCADA) A methodology involving equipment that both acquires data on an operation and provides limited to total control of equipment in response to the data.

surface water system A water system using water from a lake or stream for its supply.

surge pressure A momentary increase of water pressure in a pipeline due to a sudden change in water velocity or direction of flow.

swab Polyurethane foam plug, similar to a pig but more flexible and less durable.

telemetry A system of sending data over long distances, consisting of a transmitter, a transmission channel (wire, radio, or microwave), and a receiver. Used for remote instrumentation and control.

total dynamic head The difference in height between the hydraulic grade line (HGL) on the discharge side of the pump and the HGL on the suction side of the pump. This head is a measure of the total energy that a pump must impart to the water to move it from one point to another.

total organic carbon (TOC) The results of a general analysis performed on a water sample to determine the total organic content of the water.

total static head The total height that the pump must lift the water when moving it from one point to another. The vertical distance from the suction free water surface to the discharge free water surface.

transmission channel In a telemetry system, the wire, radio wave, fiber-optic line, or microwave beam that carries the data from the transmitter to the receiver.

transmitter In telemetry or remote instrumentation, the device that converts the signal generated by the sensor into a signal that can be sent to the receiver–indicator over the transmission channel. Also known as *transducer*.

tree system A distribution system layout that centers around a single arterial main, which decreases in size with length. Branches are taken off at right angles, with subbranches from each branch.

US Environmental Protection Agency (USEPA) A US government agency responsible for implementing federal laws designed to protect the environment. Congress has delegated implementation of the Safe Drinking Water Act to the USEPA.

variable frequency Relating to a type of telemetry signal in which the frequency of the signal varies as the parameter being monitored varies.

velocity The speed at which water moves; measured in ft/sec or m/sec.

voltage (1) A measure of electrical potential. (2) In telemetry, a type of signal in which the electromotive force (measured in volts) varies as the parameter being measured varies.

water audit A procedure that combines flow measurements and listening surveys in an attempt to give a reasonably accurate accounting of all water entering and leaving a system.

water hammer The potentially damaging slam, bang, or shudder that occurs in a pipe when a sudden change in water velocity (usually as a result of someone too-rapidly starting a pump or operating a valve) creates a great increase in water pressure.

water horsepower (WHP) The portion of the power delivered to a pump that is actually used to lift water. Compare with *brake horsepower* and *motor horsepower*.

wire-to-water efficiency The ratio of the total power input (electric current expressed as *motor horsepower*) to a motor and pump assembly, to the total power output (*water horsepower*); expressed as a percent.

work The operation of a force over a specific distance.

work order A form used to communicate field information back to the main office. Also used to order materials, organize field crews, and obtain easements.

Index

NOTE: *f* indicates a figure; *t* indicates a table.

D

E

P

R